The Future of Social Epistemology

Collective Studies in Knowledge and Society

***Series Editor*:** James H. Collier is Associate Professor of Science and Technology in Society at Virginia Tech.

This is an interdisciplinary series published in collaboration with the Social Epistemology Review and Reply Collective. It addresses questions arising from understanding knowledge as constituted by, and constitutive of, existing, dynamic, and governable social relations.

The Future of Social Epistemology: A Collective Vision edited by James H. Collier
Social Epistemology and Technology: Toward Public Self-Awareness Regarding Technological Mediation edited by Frank Scalambrino
Socrates Tenured: The Institutions of 21st Century Philosophy, Adam Briggle and Robert Frodeman
Social Epistemology and Epistemic Agency edited by Patrick J. Reider

The Future of Social Epistemology

A Collective Vision

Edited by
James H. Collier

ROWMAN &
LITTLEFIELD
INTERNATIONAL

London • New York

Published by Rowman & Littlefield International, Ltd.
Unit A, Whitacre Mews, 26–34 Stannary Street, London SE11 4AB
www.rowmaninternational.com

Rowman & Littlefield International, Ltd. is an affiliate of Rowman & Littlefield
4501 Forbes Boulevard, Suite 200, Lanham, Maryland 20706, USA
With additional offices in Boulder, New York, Toronto (Canada), and Plymouth (UK)
www.rowman.com

British Library Cataloguing in Publication Data
A catalogue record for this book is available from the British Library

ISBN: HB 978-1-78348-265-8
PB 978-1-78348-266-5

Library of Congress Cataloging-in-Publication Data
The future of social epistemology : a collective vision / edited by James H. Collier.
pages cm. — (Collective studies in knowledge and society)
Includes bibliographical references and index.
ISBN 978-1-78348-265-8 (cloth : alk. paper) — ISBN 978-1-78348-266-5
(pbk. : alk. paper) — ISBN 978-1-78348-267-2 (electronic)
1. Social epistemology. 2. Knowledge, Theory of. 3. Knowledge, Sociology of.
I. Collier, James H., 1961– editor.
BD175.F88 2016
121—dc23 2015028279

∞™ The paper used in this publication meets the minimum requirements of American National Standard for Information Sciences—Permanence of Paper for Printed Library Materials, ANSI/NISO Z39.48-1992.

Printed in the United States of America

Contents

Introduction

James H. Collier

The Future of Social Epistemology: A Collective Vision arises from a collaborative effort, made by an allied group of scholars representing fields across the humanities and social sciences, to explore questions regarding the social nature of knowledge. A common fascination with knowledge as a social phenomenon unites the contributors. At the same time, we vary widely in our conceptions of what constitutes social epistemology, and of where and with whom its future resides. We appear, on Isaiah Berlin's rendering, rather more fox than hedgehog.

We approach social epistemology as a shared normative project with two primary dimensions: (1) we take the social formation of knowledge and belief as an object of study and (2) we experiment with *doing* social epistemology as a collective practice. We focus on actively pursuing the means, aims, and ends of conceiving, developing, and distributing knowledge. We tend not to take a reductive view of epistemology as consigned to deduce reliable processes for arriving at the truth. Rather, we seek the comprehensive advantages afforded by interdisciplinary inquiry and a desire to reimagine fruitfully one's philosophical commitments given empirical findings and practical results. A practical result of our approach follows on these pages.

STORY AND STRUCTURE

Launched in November 2011, the Social Epistemology Review and Reply Collective (SERRC; social-epistemology.com) serves as both a digital wing for the journal *Social Epistemology* and as an independent platform for scholarship, commentary, and judgment on issues related to knowledge, culture, and policy. As a digital wing of *Social Epistemology*, the SERRC encourages extended dialogue on published articles, conducts interviews with authors, and produces roundtable book reviews. As a platform for experiments in and about social epistemology, SERRC members initiate and develop projects that take advantage of the capacity for online media to sustain a dialogue on complex ideas and evolving arguments.

In September 2013, SERRC members began contributing to the new "Collective Vision" page. The page was created as a forum not only for "redressing fundamental philosophical problems pertaining to knowledge, science, truth, belief, normativity and judgment, but also for imagining and creating a shared future by forwarding ideas regarding humanity, technology, embodiment, morality and governance" (social-epistemology.com/collective-vision). The entries take up a range of topics from visioneering, the future role of intellectuals, conceptions of reason and transhumanism to collective judgment, critical theory, and public philosophy. While distinctive, each entry relayed a concern with how we might consider the future—the future agenda and conduct of the field of social epistemology.

The Collective Vision page served as the starting point for this volume. SERRC members posted pieces to the page on a weekly basis. As the number of pieces grew, the dialogue among the contributors deepened. As contributors raised new arguments and topics, patterns emerged regarding our interests and concerns. We realized that, taken together and refined, the pieces fit into a fascinating mosaic. Revisiting their work, the contributors rethought, revised, and expanded their pieces into book chapters.

PART I OF THE BOOK

Our book is divided into four parts. Part I, "Conducting Social Epistemology," addresses issues regarding the purposes of social epistemology and the possibilities of realizing a "collective vision." Fred D'Agostino's opening chapter sets the tone by retracing the dynamics of the contributions to the Collective Vision web page. These contributions reflect a tension in social epistemology. While social epistemology is a recognized field possessing the requisite markers—conferences, journals, handbooks, a book series, and a named chair—it exists either as a well-rehearsed, ordered colloquy following the philosophical tradition of Cartesian epistemology, or as a contested space of communal inquiry promoting a progressive sense of what knowledge is and how it should be pursued. The contested space of social epistemology offers the creative tension of both risk—lack of position, recognition, and predictable outcome—and reward—greater freedom to explore, diversity, and vibrancy. D'Agostino acknowledges this tension in calling for an imaginative rendering of common standards.

Joan Leach's chapter embraces the chance to examine alternative ways—comics in this case—to distribute knowledge. Seeing the virtue in disagreement unless called to do otherwise, Francis Remedios bets the relative success of a collective vision on realizing Steve Fuller's conception of Humanity 2.0. David Budtz Pedersen proposes an application for social epistemology in determining what counts as research excellence. A similar concern for social epistemology's purpose comes in Inanna Hamati-Ataya and Stephen Norrie's linked contribution. Both of them argue that social epistemology's reflexive attention to the normative rationale of its knowledge practices might well serve to model what a sociological self-consciousnessness can achieve. Part I's final chapter, coauthored by Susan Dieleman, María Navarro, and Elisabeth Simbürger, analyzes how a truly "*sociological* social epistemology" might locate and bring about its political commitments, at least in part, by reconceiving familiar sites, and engaging unfamiliar sites, where we produce knowledge.

PART II OF THE BOOK

In part II, "Extending Conceptions of Knowing," we look at various ways to express and augment how and what we know. As Martin Evenden observes, to extend knowledge we must understand how metaphors affect the public perception of science and, thus, the possibilities for public debate and policy. Gregory Sandstrom seeks to curtail, if not end, memetics theorizing, associated with Richard Dawkins's research, and deploy the notion of human extension—ways in which human beings realize themselves, and their knowledge, as part of enhanced and increased mental and physical frameworks (as readily found in digital media).

Advocating a dialectical process of extending knowledge, Patrick Reider holds that we account for our private and cultural desires as we decide how best to pursue knowledge. Similarly, a dialectical process by which we determine epistemic conduct and values—the most prominent example being, perhaps, objectivity—animates Guy Axtell's contribution to our book. Philip Olson expands the contexts in which we pay attention to the qualities of epistemology by responding to research on ignorance. Olson adds the complication of whether we may wish to become "oppositionally ignorant" in situations where the resulting knowledge—from a genetic test, for example—can be used by ourselves, and others, toward unjust ends. Adam Riggio's coda to part II lends a personal perspective on how universities, in particular, take advantage of scholars' dedication to creating knowledge though the current, exploitive system of under-employing and compensating qualified scholars. Undaunted, Riggio locates alternative outlets for expressing critical knowledge.

PART III OF THE BOOK

Part III of *The Future of Social Epistemology* speaks to an issue essential in conceptualizing knowledge and, thus, pursuing relevant research—the relationship between the individual and the collective. Probing this relationship lies at the heart of assembling a "collective vision." Eric Kerr questions the very idea of what we mean when we say a group has knowledge, a vision or visions (plural). Clearly we, in the SERRC, influence one another. How, then, should we avoid the seduction of groupthink? Ought we avoid altogether the call for a collective vision? Moreover, how should we account for disagreement? Jonathan Matheson explores the Equal Weight View (EWV)—one should give the opinions of advisors that count as epistemic peers' equal weight—in light of group inquiry. Matheson challenges EWV in order to clarify what, if any, skeptical consequences follow from its prescriptions. In a similar philosophical key, Mark West encourages us to go about the work of grounding social epistemology's theoretical perspectives. In a fascinating example, West investigates the "use of symbolic language and social support" by Alcoholics Anonymous as a way to show how the practice of social epistemology produces a demonstrable empirical effect on human behavior. Miika Vähämaa, in addition, finds that a focus on social forces as "senses" underscores both the affective and intellectual dimensions regarding how knowledge gets circulated in social groups.

One social group in particular, scientists, as Diana Rishani reminds us, need to be recognized as promulgating values brought to light by feminist standpoint theory. Far from the thin version of free-floating objectivity referred to in many discussions of science, feminist theorists push us to have a fuller sense of objectivity that accounts for human perspectives often marginalized or ignored. Yet, attaining and knowing even a more fully realized sense of objectivity, for example, comes at a cost. Fabien Medvecky concentrates on the cost of communicating and disseminating scientific knowledge. Medvecky trades, in part, on Miranda Fricker's concept of epistemic injustice, to propose a basis on which we communicate justly and, as a result, distribute fairly resources such as credibility. Rather more starkly, Pedro Saez Williams claims that the future of social epistemology resides with "furthering its commitments to naturalism." Williams explains that further commitments to naturalism entail that social epistemologists, in maintaining their normative stance, must uphold a naturalism that unceasingly scrutinizes and prevents knowledge from becoming a taken-for-granted vehicle of power.

PART IV OF THE BOOK

"Envisioning Our Human Future," part IV of the book, reflects social epistemology's specific concern with the power of science and technology to alter human minds and bodies. Laura Cabrera, William Davis, and Melissa Orozco wrote their chapter as a way to try to bring together their three perspectives about technological futures in a common vision. Their chapter illustrates the process of "visioneering," and performs an experiment in collaborative authorship. Emma Craddock takes the Zapatista Army of National Liberation as an example of how we can target our imaginations to achieve great ends without being victimized by grand narratives. Our imaginations might well extend beyond the limits imposed by embodiment to artistic expressions of human augmentation and enhancement. By eliciting viewer responses to her exhibited painting, Victoria Peake sought to gauge opinions on human enhancement. The worry not only about what human enhancement would socially demand, but also about who, in an increasingly inequitable world, might benefit from such potential improvements, signals that we have much work to do in imagining where, and how, we wish to rapidly transform ourselves. Still, the characteristics that uniquely indicate what makes us human remain largely disputed.

Looking at Donna Haraway's concept of the cyborg, as framed and presented by the feminist project, Steve Fuller and Veronika Lipinska examine gender identity in our trans- or post-human future. Lipinska, interviewing Natasha Vita-More and Rachel Armstrong, two prominent women in the transhuman movement, surveys numerous possibilities for how we might identify ourselves, including as "infomorphs," as we move beyond the female/male binary. And as Emilie Whitaker points out other binaries, such as the left/right divide in politics, seem relics in the wake of visioneering through generational shifts. In speaking with a group of seventy-five children aged twelve to sixteen, Whitaker found extraordinarily complex attitudes toward the mixing of biology, identity, and technology that require an "epistemological humility" as we envision and design the future. The complexities of the future will, at the very least, severely test the Enlightenment concepts and frameworks—such as justice, the individual, the nation-state—on which we delimit notions as to what counts as "mine" and "yours." Steve Fuller tackles the thorny problem of what becomes of concepts like individual rights as we, with increasing dispatch, reconceive and reconfigure the traditional social bonds on which those concepts depend. If, for example, we continue to go enthusiastically about the business of transforming and extending the boundaries of the human body and mind, what then of the idea of an individual, and individual rights, if one may lay claim to more than one mind and more than one body?

Taken together, the book's four parts demonstrate an ongoing collaborative experiment in effecting social epistemology. The chapters deal, in various lively forms, with how we, as essentially social beings, form knowledge and belief. The variety of topics and practices found in these chapters points both to the viability and vitality of the social epistemology's future. While we offer a collective vision, we do not envision only one horizon.

Part I

CONDUCTING SOCIAL EPISTEMOLOGY

Chapter 1

How Can We Collectivize a Set of Visions about Social Epistemology?

Fred D'Agostino

As readers will be aware, this volume results from an earlier, online encounter involving participants from a range of disciplines and tackling a range of topics that have come to be associated with the idea of social epistemology. It was an important feature of this earlier encounter that the individual contributions to it appeared more or less one by one over an extended period of time so that, in theory at least, each contributor, except for the first few, had before her the contributions that had already been made. Notwithstanding this situation, almost no one referred, in her own contribution, to any of the contributions that had already been posted. This is understandable, given the pressures of time, but it was also very striking, especially in a project organized around the central ideas of social epistemology which, not to trivialize its many facets, is certainly devoted to the idea that learning is unavoidably and indeed thankfully a social activity, so that what one person says can correct or amplify or contextualize what another says. Very little of that happened in the earlier encounter.

DIVERSITY AND COLLABORATION

There is, of course, an "upside" to the fact that each contributor developed his own ideas more or less without reference to the thinking of others who, in the online encounter anyway, posted before him. A *diversity* of thinking is certainly important in the development of *better* thinking—that is fundamental. But responding to what others have said, rather than "saying your piece," can inhibit the diversity of thinking through a variety of familiar mechanisms. Certainly, fewer different things will get said if each participant does respond to the contributions of earlier participants. In this situation, what the earlier participants "put on the table" becomes an agenda for others following them, so that, unless what they might have wanted, antecedently, to say fits that agenda, they may not be able to say it (without, anyway, "changing the subject").

Of course, when contributors *do* respond to what others have already said, we are likelier to get a connected and "thematic" discourse that may be easier for the outsider, for instance, to follow. Indeed, such a connected discourse may give contributors themselves a sense that they all belong to a discursive community in which what each says gets recognized as a contribution to an ongoing discussion and perhaps even valued by the others.

Nevertheless, as Navarro (2013, 27) puts it, being responsive to what's already been said can lead to "a herding effect" which can "produce suboptimal . . . outcomes." We all end up talking about a few points, not to be distinguished in relation to their importance and potential relevance to collective objectives from others not discussed, except in relation to their order of introduction into the discussion. Precisely the points that, had they been made, might have carried the discussion to an "optimal outcome" may never get made or, even if made, will not be heard as contributions insofar as they depart from an evolving "script" for the discussion.

There is, then, something to be celebrated in the way in which participants in the earlier, online discussion "got their two cents in" more or less without reference to what others had already said. Diversity has been preserved, not inhibited. Nevertheless, now that a variety of topics and concepts *are* on the table, we can at least begin to put them in contact with one another in order to build up a picture of their relations and indeed of their utility, when taken together, in promoting the growth of knowledge.

Again, where to start is potentially quite arbitrary. While we want to put themes that were developed independently into contact with one another, we do not want to let that produce another kind of "herding effect." Certainly, I am more interested in some of the many topics introduced in this collection than in others. I will look for a strategic entering wedge, but others would see the wedge somewhere else entirely. My attempt at a (partial) synthesis is meant, then, as an *illustration* of how we can put things together to develop something that amplifies, corrects, and nuances. It is not meant to preempt others' attempts to do the same sort of thing but in ways that reflect their interests and concerns.

ASSEMBLING OUR IDEAS

Here, then, is an example of how we might start to put ideas together. It reflects my own current interest in the disciplines and how they work to secure the engagement of different individuals in common projects (for earlier work on this project, see D'Agostino 2012). In his contribution, Patrick Reider (2014) refers to people's tendency "to refrain from investing significant resources into projects with no accepted criterion of return" (54). This is certainly familiar in the literature about the scholarly disciplines. Each discipline, on some accounts, defines a standard list of problems or issues of strategic importance and constitutes, inter alia, a reward system for scholars addressing those issues (see, for instance, Whitley 1984). Since a "criterion of return" depends, within the scholarly community as much as within the market for goods and services, on what other individuals are willing to recognize as valuable, Reider quite properly concludes that "the perceived good of knowledge acquisition and production cannot be a pure expression of a solitary individual" (54). This is a nice point and, I think, is properly characterized as an ontological point. If the criterion of return in a given domain is a social artifact—a solitary individual does not get, on her own, to determine how the community will recognize and value her contribution—then, since the growth of knowledge depends, *inter alia*, on how knowledge-making activity is rewarded, the growth of knowledge is a matter that is mediated socially, just as social epistemology has insisted (see, of course, Fuller 1988, 3).

Having established the ontological grounds of a genuinely social epistemology, Reider also notes, however, that "when knowledge production and acquisition have no clear sense of benefit (material, spiritual, physical, ethical, etc.) or privilege, the degree to which a community is willing to invest in it diminishes" (54). This point is also a good one and it is a methodological point, rather than an ontological one. What it says, and again this is recognized in the literature about the disciplines, is that you cannot expect scholars to conduct research on topics or using tools that are not valued within their communities of inquiry. If they do, they cannot expect a "return on their investment" in the form of the kind of recognition and reward that will be distributed to others working within the mainstream. This point too, though in quite a different way, is relevant to understanding the project of social epistemology.

To see this we need, however, another point, one made explicitly by Melissa Orozco—namely, that "the emergence of this new interdisciplinary project [of social epistemology], in any existing version, was not friendly to existing approaches to knowledge" (17).

We now have two premises of an argument that explains something about social epistemology. Here goes.

1. The criterion of return on investment is defined with respect to the recognized approaches within a given domain of inquiry.

2. Social epistemology situates itself outside those approaches that are recognized within the domain, specifically, of epistemology as a subdiscipline of philosophy.

Accordingly it will be risky, at best, to invest one's energies in prosecuting the project of social epistemology; there is no community whose criterion of return allocates recognition and reward to the practices that are constitutive of social epistemology.

Put crudely and with greater generality, scholars pursuing a new, unfamiliar, only partly institutionalized research agenda will not be able to present their findings for evaluation against already established criteria of return and hence risk not getting the sorts of returns that sustain commitment to the project. Notice the resonance of this point to the slightly different, but certainly related one of Stephen Norrie (2013) in his acute analysis of the way in which critical theory "is completely incompatible with existing academic practice" (32) (and its criteria of return), and of Inanna Hamati-Ataya (2014) in her point about the inconceivability "outside the SE community" of its "premises . . . constitutive theoretical traditions and the empirical literature it draws on and produces" (44). It is also, I think, alluded to by Elisabeth Simbürger (2014) in her reference to "the labour relations of academic knowledge production" (5). Insofar as social epistemology, or indeed *any other "revolutionary" approach*, rejects established projects and their criteria of return, it becomes a risky enterprise for individuals to engage in who have careers to make which *can* be made, at least typically, only through the mechanisms of recognition and reward that they have distanced themselves from.

On the other hand, as Orozco (2014) also points out, the separateness of social epistemology (from more established academic programs) might facilitate "its creative drive and quest for a vision," an idea related, I think, to Jonathan Matheson's (2014) point about the importance, in avoiding "confirmation bias," in a group dynamics which (like the one we are attempting here) not only suffers, but honors, the participation of dissenting members, or, anyway, members with a diversity of interests and approaches. Again, the reasoning is familiar. If we are willing to forego recognition and reward, we need not narrow our range of interests and concerns (to fit the disciplinary template). But this means that our diversity of interests and concerns remains available to us as individuals and, more importantly, to our collaborators and colleagues. And this means that all the potential for correction, amplification, and nuance that is implicit in that diversity remains available to us as we form and refine our views. We do not inhabit an echo chamber in which each of us is singing from the same sheet. We may not know, with the sort of certainty and confidence of the fully disciplined scholar, how to allocate recognition and reward, but we do not have to wonder, we can hear, what scholars with different views have to tell us; they will not be inhibited from telling us out of fear of not being rewarded for their orthodoxy. Of course, they won't be as reliably rewarded anyway. That is what follows from there not being an orthodoxy. Rewards come, or not, depending on the reactions of individuals.

With a well-established research agenda, we have a more or less mechanical allocation of recognition and rewards. Without such an agenda, recognition and reward are haphazard and unreliably delivered on an ad hoc basis. With an established agenda, we risk "conformism" in thought, with all that implies. Without such an agenda, we have the potential for the creative combination of ideas and techniques that do not already "go together" or apply (as seen from more conventional standpoints). Social epistemology explains all this when it looks at the production of knowledge in the various disciplines. But, as a scholarly formation, it also experiences all this: it is creative, but recognition and rewards are not delivered mechanically, and hence it is a risky enterprise.

This issue is crucial for all knowledge production. How do we manage or, better still, "tune" the tension between reliable distribution of rewards (which seems to require set standards and hence an orthodoxy) and the conditions required for genuine creativity (which seem to demand diversity in thinking and action and hence a heterodoxy).

Certainly, to be on the side of creativity is, at least potentially and probably typically, to put some distance between oneself and the system of routine recognition and reward. It therefore requires courage or a bias toward risk taking to pursue this path. But it will also require, I submit, the development of institutional forms, perhaps sitting at an angle to what Norrie refers to as existing "trajectories of career advancement" that take up the challenge of providing both a sympathetic audience and one that is also capable of critical engagement in support of standards of excellence.

Let me take these points in order.

It is, as the remarks of Reider remind us, psychologically harder to pursue a path that has not already been well trodden by others to the extent, perhaps, that it has been "domesticated," stripped of its founding excitements and risks. For this we need recognition but can't reliably get it, in many cases, from our immediate colleagues in the offices down the corridor from us. So we need to support each other in our joint and risky venture.

Secondly, however, we need to make up as we go precisely the criteria that will enable us to distinguish between good and better and, indeed, not so good work within this emerging landscape of ideas and methods. If we want social epistemology, or indeed any other revolution in scholarship, to become more influential in any of the various ways that might be important, then we will have to develop and enforce our own standards of excellence. They may differ in various particulars from the standards current in some of our original disciplines, but in certain crucial respects they will need to remain recognizable from within those disciplines by those with, anyway, their own forms of courage and risk taking, their own forms of openness to what is not conventionally preapproved.

Here, I think, Mark Douglas West's (2014) contribution holds a lot of promise. What West outlines is an approach to changing ourselves and to risk taking in aid of changing ourselves, and to mutual sympathy in support of risk taking, and nevertheless to demanding standards in relation to self-change that has proven successful in the extremely challenging circumstances associated with addiction. Translating West's account of Alcoholics Anonymous to our own current circumstances, the story might go something like this. Each of us has to acknowledge something about the inadequacies of our previous practice in relation to our deepest aspirations for understanding knowledge and/or for applying knowledge in aid of more effective democratic practices. Each of us has to be willing to take the risk of acknowledging the importance to a possible future practice of our openness to the kinds of ideas and methods reported in this Collective Vision statement and elsewhere in the social epistemology literature. Each of us needs to acknowledge that we need the support of others and are able and willing to lend it to others who are also in need. And each of us needs to honor the imperatives associated with this collective. As Thomas Basbøll (2014) put it, "We have to become certain kinds of people in order to know certain kinds of things," (14) or, perhaps, we have to become different kinds of people in order to know things in different kinds of ways, in ways that are transparently collectively mediated.

I said earlier that some of the inhibitions or disincentives to the creation of new forms of and approaches to knowledge (and its use) might be related to our disciplinary situations. I want, in conclusion, to return to this point.

I participated, a while ago, in a septennial review of a School of Tourism (Studies). The external reviewers were distinguished international figures in tourism studies. Nevertheless, they were concerned about the (so far only partial) disciplinarization of tourism studies as such. Each of them had come to tourism as an object of inquiry from an already established and topically broader field of inquiry—one from geography, one from anthropology, and one from economics. These founding generation tourism studies academics each had a background in an already-existing discipline and they were, collectively, worried that the *next* generation would all have come from *within* the "field" of tourism studies, and thus would lack a full understanding of the methodological tools and theoretical and conceptual apparatus that a more established discipline can provide.

Of course, in establishing tourism studies as a "sort-of" discipline, these founders had difficulties, in the unavoidably multidisciplinary context of early tourism studies programs, in making themselves understood to one another, in developing criteria for the assessment of their work that could be "trusted" by a collective from a variety of different backgrounds and so on. But they did do this. That's the take-home message. I think it may have been easier to do this because, or to the extent that, they were able to achieve a degree of institutional and institutionally recognized autonomy—for example, their own conferences, journals, eventually academic departments, and so on. Perhaps this suggests something about the conditions for the possible success of the social epistemology project, as Hamati-Ataya might put it. Certainly, there are already conferences, journals, and, as this volume itself shows, publishers' commitments. The prospect, and perhaps even the desirability, of a separate organizational unit (e.g., a department) is less obvious, I think. The recent history of interdisciplinary institutional configurations is certainly mixed at best and putting together sociologists, philosophers, cultural studies practitioners, and others in a single department runs another kind of risk—one that the postgraduate students in such a department would have to bear. Typically, single-discipline departments reproduce themselves by graduating students who will be recognizable to those hiring in other departments that are organized on the same basis. But a department of social epistemology will be, at least pro tem, an oddity in the academic world and hence there will be, again pro tem, no obvious "placement" opportunities for its graduates.

I think that "brand" has something to do with the possible conditions of success for such a project. Some of the institutionally and even intellectually more successful attempts at collective and interdisciplinary new approaches to knowledge production have "brands" that work pretty readily to draw attention—for example, gender studies, American studies, early modern studies. Because they refer to a topic that people from various disciplines might already have an interest in and competence to pursue, they recruit these people, alerting them to the possibility of fruitful collaboration around a shared interest. I am not sure, and I say this with the profoundest respect for much work that already flies under the banner of "social epistemology," that those two words are working as well as some of these other labels have. Unfortunately, I have nothing better to suggest. I will, however, keep thinking about that.

The essential problem for us is to find and gradually institutionalize ways to recruit, to recognize for, and to create some common standards for judging work in "social epistemology." In doing that, we have to honor the tension that sits underneath this objective as, indeed, it does under every project of knowledge making. This is the tension between, on the one hand, having a reliable mechanism for distributing recognition and reward for individuals' contributions to collective knowledge making and, on the other hand, retaining the sorts of diversity of thinking that facilitate creativity. We have made a start.

BIBLIOGRAPHY

Basbøll, Thomas. "I, Social Epistemologist." *Social Epistemology Review and Reply Collective* 3, no. 4 (2014): 14–15. wp.me/P1Bfg0-Y9.

D'Agostino, Fred. *Naturalizing Epistemology: Thomas Kuhn and the "Essential Tension."* Basingstoke: Palgrave Macmillian, 2010.

D'Agostino, Fred. "Disciplinarity and the Growth of Knowledge." *Social Epistemology* 26, no. 3–4 (2012): 331–50.

Fuller, Steve. *Social Epistemology*. Bloomington: Indiana University Press, 1988.

Hamati-Ataya, Inanna. "Before Social Epistemology: On the Limited Efficacy of 'The Scandal.'" *Social Epistemology Review and Reply Collective* 3, no. 2 (2014): 44–48. wp.me/P1Bfg0-Y9.

Matheson, Jonathan. "A Puzzle about Disagreement and Rationality." *Social Epistemology Review and Reply Collective* 3, no. 4 (2014): 1–3. wp.me/P1Bfg0-Y9.

Navarro, María G. "How to interpret collective aggregated judgments?" *Social Epistemology Review and Reply Collective* 2, no. 11 (2013): 27–28. wp.me/P1Bfg0-Y9.

Norrie, Stephen. "Can Critical Theory be Computerised and Collectivised?" *Social Epistemology Review and Reply Collective* 2, no. 12 (2013): 32–36. wp.me/P1Bfg0-Y9.

Orozco, Melissa. "Visioneering Ourselves as Social Epistemologists." *Social Epistemology Review and Reply Collective* 3, no. 7 (2014): 16–18. wp.me/P1Bfg0-Y9.

Reider, Patrick J. "Knowledge as a Public Commodity." *Social Epistemology Review and Reply Collective* 3, no. 6 (2014): 53–55. wp.me/P1Bfg0-Y9.

Simbürger, Elisabeth. "Social Epistemology as Work: A Quest for Normativity at the University." *Social Epistemology Review and Reply Collective* 3, no. 4 (2014): 4–6. wp.me/P1Bfg0-Y9.

West, Mark Douglas. "Doxastic Involuntarism, Attentional Voluntarism, and Social Epistemology." *Social Epistemology Review and Reply Collective* 3, no. 5 (2014): 37–51. wp.me/P1Bfg0-Y9.

Whitley, Richard. *The Intellectual and Social Organization of the Sciences*. Oxford: Clarendon Press, 1984.

Chapter 2

A Comic Moment for Social Epistemology

Joan Leach

> Once [a] book on Newtonian mechanics is put in the hands of many different people, each of whom has read many other different books, it is simply bad psychology to expect that this multitude of readers will uniformly use the book as intended. Maybe these books on Newtonian mechanics will succeed at bringing people closer to the nature of physical reality, but maybe not, and maybe (and most likely) they will bring most of the people closer but at the cost of making some other aspect of reality more mysterious than before. In any case, the exact outcome of this situation—how the particular distribution of texts will end up affecting the overall social order—is far from clear. However, for someone interested in social epistemology, this is the real question about the nature of knowledge that all the other questions are trying to pose.[1]

What are the appropriate objects of social epistemological study? Or, more provocatively, what can a comic strip—as opposed to a book on Newtonian mechanics—tell us about the distribution of knowledge, especially technical knowledge emerging from Big Science projects toward the end of the twentieth century? The comic example of this essay reinvigorates the motivating question for social epistemology—how does knowledge distribution affect the social order? More specifically, what sorts of knowledge distribution activities are methodologically appropriate for social epistemology? By following an epistemic artifact—a comic—from the outside-in, this essay will point to the ways in which social epistemology reveals the various knowledge stories that inform thinking about knowledge dissemination.

From 1961 to 1979 an Australian science "faction" comic called *Frontiers of Science* was syndicated to around two hundred newspapers in nineteen

Figure 2.1 ***Frontiers of Science*** **on Relativity**

languages around the world. In the 1970s when the comic was in daily papers in Australia, it easily had a reach of three million people on any given weekday.[2] What is remarkable about this, in addition to its reach, is the way that it was used. Children and teachers would cut the strips from the comic pages of the newspaper (where it might appear under *The Batman*) and make their own ersatz textbooks for use in the classroom. They also made personal albums and collections of favorite strips that had acquired personal significance for individual readers. The strip also motivated questions and critique from its readers about the nature of science, knowledge, and issues of ethics and equity in the way science was then practiced. Indeed, readers would write into newspapers, as one woman did to the *Toronto Star*, complaining about how women were not represented in the comic, and, by association, in scientific research.[3] The strip is also remarkable in the way that it framed the issues of the 1960s that would become the wicked problems of the end of the century. The comic suggested in 1961, for example, that solving the problems of water shortage might be as important as getting a man to the moon.

The comic further argued that technofixes to pollution, overpopulation, and energy shortage could be found in dustbin rocketry, birth control pills, and everything nuclear. In this way, the comic framed emerging social and technical problems as scientific and sold readers on the technofixes that were drawn to look possible and achievable with a bit of scientific fortitude. In short, *Frontiers of Science* is a remarkable artifact of knowledge dissemination; it appears at first glance to fit squarely in the traditions of science popularization, if more successful than most attempts. It was successful in the sense

Figure 2.2 ***Frontiers of Science*** **on Water Policy**

Figure 2.3 ***Frontiers of Science*** **on Probability Theory**

that it reached many people and that it was used as a vehicle of disseminating scientific knowledge to a wide and varied readership. But, like textbooks and other more elite knowledge vehicles, the knowledge it shared was framed in complex ideological formations. So, it was possible to read an accessible comic explanation of probability that would match most textbook versions but also be asked to accept that probability defined the future advancement of science and industry. The comic thus poses a series of important questions for those who are interested in the successful communication of scientific knowledge and those questions have a distinctly social epistemological flavor.

Our comic guide for the week is the Chevalier de Méré, the gambler who is said to have motivated the mathematics of Pascal and Fermat. After working through his gambler's problems and introducing textbook approaches to probability, this strip proposes that these basic questions that the reader now understands will motivate new developments in travel and energy—the Chevalier appears in front of a nuclear reactor.

Social epistemology is well poised to become more expansive in the subjects, objects, and stances it regards as relevant to epistemology. Naomi Scheman, for example, has argued persuasively about the absurdity of mapping epistemic categories such as credibility directly on to hierarchies of power. She writes,

> Although vulnerability can be an epistemic advantage, when it comes to credibility in the eyes of the powerful, it tends to be a liability, and a contagious one. Taking the vulnerable seriously, being willing and able to learn from them, is not in general a good way to bolster one's own credibility. In particular, those whose jobs put them in such a position tend not to be listened to by those in positions of greater power. Thus, for example, nurses are in a position to learn from patients things about the courses of their illnesses or injuries that escape the official measurements, but doctors do not typically take nurses to be credible sources of genuine knowledge.[4]

The stance of epistemic agents, then, is one where Scheman and others, including epistemologists of ignorance, argue that social epistemology would be well placed to give attention.[5] The subjects of epistemology, too, go hand-in-hand with this proposal. While experts and elite knowledge holders have traditionally been of interest, if one of the normative goals of epistemology is to disseminate the best knowledge possible, surely it is not only the knowledge holders, but the coproducers of knowledge and those who make sense of the knowledge as it is given. Finally, in addition to subjects and epistemic stance is an analysis of knowledge itself, which inconveniently for some, comes embodied or incapsulated or textualized or otherwise mediated in the forms in which it is presented. For social epistemologists, scientific knowledge has been a focus, but usually scientific knowledge as it is produced and disseminated (and received) among knowledge elites.

From these challenges of broadening the focus of epistemology to the contexts of knowledge production and dissemination comes a range of methodological issues which themselves are worthy of study. Following Fuller (2012), there seems to be a robust distinction between epistemology from the "inside-out" and epistemology from the "outside-in."[6] Starting from the inside, which usually is the epistemic state of an individual knower, the social epistemologist pays special attention to knowledge practices that achieve particularly good or poor results in light of the particular virtue at which they aim. Starting from the outside, the social epistemologist notices knowledge at work in the world and then investigates how it was produced, what effects it has and what lessons there are for future knowledge production. One aspect of starting from the outside is that the social epistemologist has an enormous potential range of knowledge artifacts to study and can potentially draw empirical evidence about those artifacts' travel from a range of disciplinary and methodological frameworks. This is not to say that the state or behavior of an individual knower is never relevant, only that it is one part of a larger story about how knowledge is disseminated.

One implication of drawing on a wider range of epistemic artifacts is that the social epistemologist may have to query their assumptions about the status of those artifacts. In this case, a comic strip would seem to have not enjoyed a lengthy history of epistemic analysis.[7] Yet, an artifact whose purpose is to convey knowledge that has an audience of several million worldwide would seem prima facie to be at least as much of interest as the book on Newtonian mechanics that begins this chapter. One worry might be about the status of the knowledge that a comic disseminates—an assumption might be that the book on Newtonian mechanics might somehow have more epistemic potential (as measured historically by the change in the state of knowledge of readers). And yet *Frontiers of Science* (and by association quite a bit of popular science) realizes this potential as well, and in addition, can have broader epistemic impacts by increasing readers knowledge about historical matters, engaging their imaginations on future impacts of science and technology, and extending readers awareness of the connections among various forms of scientific knowledge. A book on Newtonian mechanics might struggle to meet the particular epistemic gains of a comic book, if those are the gains that are valued.

Of course, the notion of a larger epistemic story, and not just the internal belief state of an individual knower, is itself of interest. Methodological questions and issues of the appropriate objects, subjects, and stances for study themselves take part in a range of knowledge stories—a set of narratives that presuppose how knowledge fits into a pattern of dissemination. A concrete example of this is the set of knowledge stories that surround science popularization—the intentional dissemination of scientific knowledge. In thinking about the category of "popular science" in the late twentieth century, there are two dominant narratives (with almost endless variation).

1. The first way of thinking is that popular science is the predominant form in which nonscientists get scientific knowledge (with the TV and now the Internet as key to consumption patterns) and this is important because popularization can be harnessed to improve levels of knowledge about science and even provide a critical "fourth estate" for science.
2. The second way of thinking is that popular science is the predominant form in which nonscientists get scientific information and this is worrying because this is where scientific knowledge goes through the mangle of popularization and is "framed," "spun," and "sold."

As Stephen Hilgartner chronicled in his canonical article on popular science, if scientists themselves get burned by the first way of thinking, they can always try the second.[8] Or, more seriously, Hilgartner reflected that these two takes on "popularization" as knowledge distribution are themselves rhetorical resources for those who will popularize to justify their activities and critique others doing the same. It is extraordinarily difficult to escape the pull of these two narratives. Both accounts are mired in wishful thinking. The first account indulges in the wishful thinking that popular forms of media will prevail where science and civic education has failed. The second account succumbs to the fantasy of an all-powerful mediation selling the dubious wonders of science to dupe after dupe.

But of course these two narratives presuppose that scientific knowledge is produced by experts and then trickled down through some set of mediations and is variously understood and interpreted by eventual readers. The example of *Frontiers of Science* points to some complications for this knowledge story. First, *Frontiers* itself was quite literally coproduced by a physicist, a documentary producer, and a cartoonist and there is substantial evidence that they negotiated the topics, the tone, the descriptions, the visuals, and the text that created the artifact. Further, the science presented is a less reported fact that science is in the making, sometimes enjoining experts and readers in a common project—to imagine what life on Venus might look like, for example. What the production of the strip reveals is that what is simply called "popularization" and placed into a knowledge story about "made facts" moving from experts to lay readers is woefully inadequate to understanding how knowledge is being produced and how it travels. Another key piece of information about *Frontiers of Science* also demonstrates that knowledge moves in various ways nearly simultaneously. Each topic covered by the comic strip over nineteen years had a source; sometimes that source was a book, like a book on Newton's mechanics, but sometimes the source was a conversation among researchers at a conference, and sometimes it came from a previous example of popularization that had already been published. A "trickle-down" knowledge story does not, then, adequately reflect the way in which knowledge moves in relation to this popular artifact.

Does this generalize to other so-called popularizations? Without extensive empirical work, this is hard to answer. But, from existing studies of "popularization," the answer is at least "sometimes."[9] Most striking in this regard is James Secord's work on the "reception" of Darwinian ideas; "reception" being the way that this work is frequently discussed in academic circles. Yet, "reception" is not highly reflective of what seems to have happened in Secord's telling. "Reception studies" become another way of locking our ideas about knowledge dissemination into well-trodden knowledge stories. His work makes it clear that the processes of sharing, adapting, interpreting, and explicating what has now become known as "Darwinism" incorporated what seems hopelessly inadequate to call "popularizations." Rather, what Secord tells the social epistemologist is about the ways "useful knowledge" circulated in light of some rather far-reaching epistemic goals. Science fiction is another genre discussed in relation to the ways that speculative scientific knowledge is shared. *Frontiers of Science* liked to refer to itself as "science faction" and there is a hidden genre of "science fiction" that takes its role seriously as a kind of speculative fiction based on solid factual grounding. Of course, what seems solid factual grounding also shifts, making this not only a speculative form of fiction revealing of prior scientific assumptions but also an important imaginative space where scientific thinking can be done in a more or less "unfettered" way.

For social epistemologists, the important message that this example sends for those interested in the power of science fiction to imagine futures,[10] visioneering,[11] or even in the role of various mediations in distributing knowledge[12] is that the form of mediation matters, especially in its ability to get an audience. *Frontiers of Science* fashioned itself as a way for papers to "sell more copies" by harnessing the fact that "everyone is science conscious these days." That is, the comic self-consciously produced a market for itself and for science. This is what is missing from quite a few accounts of knowledge distribution and especially science popularization—a description of the enabling conditions of mediation itself. In this example, the enabling conditions are a market where knowledge can be exchanged and a context in which audiences can be produced (pity the reader who thought she was just going to read *The Batman* and was confronted by a science comic instead!). This is a moment where the view of a social epistemologist can shed light on recent history and use it to reflect on the here and now and what should happen next. It might also make us add a bit more of the empirical into an account of visioneering or imagining futures where the social epistemologist can begin to parse the various rhetorical modes of spin, framing, selling, and the ways in which audiences (even for professional science) are not found but made.

What does this argument amount to? On the one hand, this could be read as special pleading for studies of popular science and there have been a number of those asking that the history and social context of popular science get more attention from the scholarly community. For the most part, however, these pleas have been to put popularization in a more central role in the history or sociology of science, and there is some evidence that these pleas have been successful in science and technology studies (STS) and history and philosophy of science (HPS) accounts. But the argument here is slightly different and rests upon the success of these previous pleas. Popularization is not just intellectually engaging or meaningful scholarship insofar as it expands our recognition of the history or sociology of science, it is important if we think about the dissemination of knowledge and how it is possible to meet normative goals of spreading and increasing scientific knowledge across a wide population in a fairly equitable way. This framing is different; it will need new knowledge stories that do not start with individual knowers, but with knowledge artifacts that can be studied from the outside-in. This social epistemology will explore the range of knowledge stories, themselves elaborate rhetorical productions, that include how knowledge circulates rather than trickles down, bounces around in an unruly fashion instead of flows, and how the knowledge in the artifact changes from a static fact to an imaginative proposition, from the fact of life on Venus to the speculation of being alone in our social system. This mode of study also has the potential to turn around the tendency to inadequate reflection about categories of "credibility" and open up what Scheman has rightly drawn to our attention; if we want to know more about the dissemination of knowledge, we might also want to know more about how it fails. *Frontiers of Science* was successful in media terms. Other so-called popularization fails. Frequently, it is said to fail because the citizenry is not ready, not prepared, not educated to the level where that knowledge can matter. That, too, is a knowledge story worth unpacking.

Images of *Frontiers of Science* reproduced with the permission of the rights holders Miriam Butler and Angela Raymond. The archive of images is available at frontiers.library.usyd.edu.au.

The work on *Frontiers of Science* is supported by an Australian Research Council Discovery grant awards to Joan Leach and Maureen Burns.

NOTES

1. Steve Fuller, "Social Epistemology: A Statement of Purpose." *Social Epistemology* 1, no. 1 (1987): 1–4.
2. Maureen Burns and Joan Leach, "Science as Extra Dividend: Frontiers of Science." *International Journal of Cultural Studies* 14 (2011): 531–46.
3. Ibid., 543.
4. Naomi Scheman, "Toward a Sustainable Epistemology." *Social Epistemology* 26, nos 3–4 (2012): 471–89.
5. See Susan Dieleman, review essay of "Agnotology: The Making and Unmaking of Ignorance," and "Race and Epistemologies of Ignorance." *Social Epistemology Review and Reply Collective* 1, no. 2 (2012): 11–25.
6. Steve Fuller, "Social Epistemology: A Quarter-Century Itinerary." *Social Epistemology* 26, nos 3–4 (2012): 267–83.
7. For notable exceptions, see Bassalla and Locke. George Basalla, "Pop Science: The depiction of science in popular culture." In *Science and Its Public: The Changing Relationship*, edited by Gerald Holton and William Blanpied. *Boston Studies in the Philosophy of Science*, vol. XXXIII. Boston: D. Reidel Publishing, 1976: 261–79. And Simon Locke, "Fantastically Reasonable: Ambivalence in the Representation of Science and Technology in Super-Hero Comics." *Public Understanding of Science* 14 (2005): 25–46.

8. Stephen Hilgartner, "The Dominant View of Popularization: Conceptual Problems, Political Uses." *Social Studies of Science* 20, no. 3 (1990): 519–39.

9. This is most evident in the important collection of articles in Roger Cooter and Stephen Pumfrey, "Separate Spheres and Public Places: Reflections on the History of Science Popularization and Science in Popular Culture." *History of Science* 32 (1994): 237–67, p. 239.

10. Victoria Peake, "The Value of Imagining the Trans/Posthuman." *Social Epistemology Review and Reply Collective* 2, no. 11 (2013): 13–15.

11. Laura Cabrera, "Visioneering and Our Common Future." *Social Epistemology Review and Reply Collective* 2, no. 10 (2013): 1–3.

12. Steve Fuller, "Intellectuals as both dangerous and endangered." *Social Epistemology Review and Reply Collective* 2, no. 10 (2013): 18–20.

Chapter 3

Knowing Humanity in the Social World

A Social Epistemology Collective Vision?

Francis Remedios

A vision statement shares the organization or community's goals with its stakeholders. In the statements given on the Social Epistemology Review and Reply Collective (SERRC), social epistemology's future remains largely unaddressed. Without specifying the SERRC's future, a definitive collective vision is difficult to assemble among the SERRC's diverse statements (Kerr 2013). Still, we find ideas as to how we might proceed. Fred D'Agostino (2014) suggests developing a scholarly society brand similar to interdisciplinary projects like gender studies. Perhaps an approach toward a collective vision resides with Carlo Martini's (2014) idea of keeping our critical ability to look for causes of disagreement unless there is new information to seek agreement. Steve Fuller (2006) shares this view.

In this chapter, I review key concepts found in Steve Fuller's work over the last fifteen years. I do so to claim that the conduct, and vision, of social epistemology resides with the changing conceptions of humanity, and related social and epistemic policy, developed by Fuller. Given, then, the changing boundary conditions of humanity, how does Fuller's current and future social epistemology fit into the collective vision of the SERRC?

Fuller's social epistemology offers a social naturalistic theory of science to develop a rational policy for knowledge production. As a moral project of humanity, the social sciences connect with Fuller's social epistemology since its central issue is a normative conception of how science should be organized resulting in a knowledge policy consisting of democratically chosen goals (Fuller 2006, 2009a; Remedios 2003, 2009). Fuller's *Knowledge* (2014a) distinguishes agent-oriented and object-oriented social epistemology. Agent-oriented social epistemology unifies knowledge based on subjective interests. For object-oriented social epistemology, knowledge is divided based on its objects. Fuller favors agent-oriented social epistemology. The agent-oriented social epistemologist is a knowledge policy maker who develops knowledge policy on the governance of science (Fuller 2000).

Other than Fuller's work, little discussion in the current literature of sociology of scientific knowledge (SSK), science and technology studies (STS), sociology of science, philosophy of science, epistemology of science, and analytic social epistemology analyzes the interplay between scientific knowledge and humanity. Exploring this interplay is important since scientific knowledge remains one of the most powerful forces for human improvement. Scientific knowledge based on nanosciences, synthetic biology, and computer technology is changing humanity. Here, I refer to humanity to designate the quality that makes humans distinct from nonhumans. Fuller's social epistemology explores the changing social dimensions of both scientific knowledge and humanity as well their interplay. Hence, the future of the epistemic agent has to be specified due to changing boundary conditions.

Focusing on the university, the premier site of knowledge production, as a corporate epistemic agent, Fuller evaluates its struggle with the forces of neoliberalism. The changing boundary conditions of the epistemic agent, the knower, due to threatened changes in humanity coming from nanosciences, synthetic biology, and computer technology are Fuller's concern with knowledge policy and epistemic justice in this wide range of issues.

Fuller criticizes STS for providing only descriptive accounts of science and not normative accounts. STS, especially as manifest in the Edinburgh School, demystified the account of scientific knowledge given by philosophers of science. Although STS is the discipline with which he is frequently associated, Fuller sees STS as being both relativistic, and as a discipline without a knowledge policy (Fuller 2007). In terms of his defense of humanity, Fuller responds to STS discussions initiated by Bruno Latour for removal of the distinction of humans and nonhumans both in matters of research and of policy (Barron 2003). Fuller disagrees that the distinction between humans and nonhumans should be removed because epistemic agency would be otiose.

Fuller argues that STS, as a discipline, failed to develop its own goals and has become increasingly client driven in these neoliberal times. Given neoliberalism's impact, in which clients can strongly influence how academic knowledge is produced, Fuller defends the university, which is a corporate epistemic agent, as the premier site for knowledge production for the public good. On the basis of epistemic justice, Fuller recommends the classical Humboldtian notion of the university that emphasizes both research and teaching. In this context, interdisciplinarity becomes a battleground. Fuller argues for a kind of interdisciplinarity that is regenerative of the university in which academics reach beyond their own fields, in contrast to a neoliberal version of interdisciplinarity in which academics work in teams for clients on projects. The social epistemologist, then, is the deviant interdisciplinary agent who organizes the disciplines within himself or herself like a "Techno-Goethe" (Fuller 2014b).

The changing boundary conditions of the epistemic agent is that the knower now is not the same as the knower of the future—a future in which humans can be enhanced through biotechnology, genetic engineering, and synthetic biology. Fuller's recent work (2011) brings together his discussions in other works on the foundation of the social sciences (2007) and the intelligent design debate (2008). As Fuller's social epistemology concerns the social transformation of knowledge, the exploration of the changing boundary conditions of the knower is critical. For example, with the advancement of computer and digital technology, avatars can be created and the identity of the knower extended. The interface between the knower and the world has changed because the knower can be changed through either human enhancement or avatars.

Assuming biotechnology, genetic engineering, and synthetic biology are changing humanity, what does it mean to be human? What is the distinctiveness of humanity? As humanity is the locus of the social sciences, the changing boundary conditions of biology and ideology remain important. With the welfare state as the locus of the battle between biology and ideology on humanity, Fuller defends the distinctiveness of humanity. Fuller diagnoses the problem of humanity to be a bipolar disorder between our animal nature (biology) and our transcendent nature (ideology). Are we closer to animals, as indicated by Darwinism, or are we closer to God, as indicated by Christianity? In today's terms, the positions can be portrayed to be between the poles of Peter Singer's animal liberation or Ray Kurzweil's spiritual machines.

For Fuller, humanity consists of a socially organized resistance to the natural selection and natural forces through collective projects such as Christianity, the university, and the state. Fuller views the project of humanity as the foundation of social sciences (Fuller 2006). Humanity is a collective project in which people are organized through norms and it is the task of the social scientist to provide a framework for this endeavor. Participation in large-scale projects allows humans to control or even reverse the effects of natural selection. Further, Fuller claims that classical sociologists—Durkheim, Marx, and Weber, for example—would concur with his characterization of the project of humanity.

Essential to Fuller's concept of the project of humanity is the redistribution of wealth through the state. Fuller recognizes Foucault's notion that the human sciences as a body of knowledge was created in the nineteenth century and by the twentieth century, man has died—human sciences as a body of knowledge are in question. Fuller connects humanity to transhumanism, which is the view that humanity can be enhanced or redesigned through technology. With converging technologies (biotechnology, nanotechnology, and computer technology), humanity can be transformed to an enhanced version of humanity—Humanity 2.0.

How did the project of humanity start? Fuller avers that John Duns Scotus started the project of humanity with a univocal theory of predication of God's attributes to humans while Thomas Aquinas had an equivocal or analogical theory of predication of God's attributes to humans. For Scotus, human's difference to God's is by degree, while, for Aquinas, human's difference to God is by kind. Humanity is created in the image and likeness of God. For scientists such as Bacon, Newton, and Mendel, who were Christians, doing science was participating in the mind of God. With the advance of the nanosciences, biotechnology, and genetic engineering, through which the future of life can be engineered, there have been many warnings that science is playing God. From Fuller's perspective, doing such science is to know the mind of God.

Fuller takes on Darwinism with intelligent design (ID) theory. ID is the view of the role of divine design in western science. In 2004, Fuller was an expert witness for the defense in the *Kitzmiller v. Dover Area School District* trial. The defense asserted that ID be taught as an alternative to Darwinism. The judge disagreed that ID is a science. In a controversial move, Fuller recommends the promotion of an Abrahamic theological perspective to motivate students to become scientists in the United States. He promotes Abrahamic theology's view that humans are privileged to understand and control nature, as they are created in the image and likeness of God.[1]

Two principles used to assess risk of the impact of science and technology are proactionary and precautionary principles. Though the proactionary and precautionary principles are about risk assessment, they are also about epistemic agency, which is the responsibility of the knower of the action. The proactionary principle stresses risk taking as definitive of the human condition while the precautionary principle stresses the need to conserve nature. Fuller explores the various futures open for the human condition, including ones that embrace "transhumanism" and "posthumanism." Starting with the emerging challenges posed by the so-called human enhancement sciences and technologies, Fuller has explored alternative futures under three rubrics: the ecological, the biomedical, and the cybernetic. These attempts to reengineer both our bodies and the environment require substantial redefinitions of social justice and economics productivity, all envisioned within a new political order of Welfare 2.0, which promotes risk taking.

Welfare State 1.0, which protects people from harm, follows the precautionary principle. Welfare State 2.0 calculates welfare differently. Instead of protecting individual human lives, it promotes risk taking. Fuller's Welfare State 2.0 three principles are:

1. Instead of discouraging risky experiments, people should be encouraged with compensation for victims of failed experiments.
2. Value preferences change with new technology.
3. Welfare State 1.0's social safety net is removed; instead, Welfare State 2.0 favors rewards for risky innovation (Fuller 2014c).

Fuller challenges the liberal welfare state, which includes protection of individual life. Welfare State 1.0 is influenced by Rawls's liberal notion of citizenship, which is commitment of an individual's freedom, equality of opportunity, and distribution of resources, for a redistributive welfare state. Fuller's Welfare State 2.0 proactionary welfare function calculates risk differently to get humans to embrace risk as collective flourishing rather than as protecting individual lives. Welfare State 2.0 is based on a republican theory of citizenship, which is based on participation in decision making, instead of the liberal notion of rights. Fuller points out that not all capacities in humans 2.0 are enhanced, so there is no master race. Finally, because of Welfare State 2.0's experimentation, and Fuller's view that humanity is immaterial, a computer with superintelligence can be built, which can kill humans like humans can.

Collin (2013) and Pedersen's (2013) criticisms of Humanity 2.0 point out that the kind of society we want will depend on the humans we want to create through genetic and moral manipulation—for example, the compassionate Christian, the peace-loving hippy. In addition, why should humans 2.0 redistribute welfare to humans 1.0? Given Fuller's utilitarianism combined with transhumanism, Pedersen counters that there is a lack of a firm axiological basis for normative guidance of moral action because the biological substrate of humans 2.0 would be radically different from that of humans 1.0.

I suggest that an alternative to Fuller's utilitarianism is virtue ethics, which is not an abstract moral theory such as utilitarianism or deontology, but emphasizes interrelatedness of agents and commitment to values beyond rules institutionalized by the state. For virtue ethics, morality is not a preference or constraint; it is internalized in practice based on seeking the good life in an information society. Nietzsche viewed utilitarianism as slave morality (every moral agent has to maximize the greatest good for greatest number). Though Nietzsche's view of morality is perspectival, it is compatible with virtue ethics. Virtue ethics, which is not an abstract moral theory such as utilitarianism or deontology, emphasizes interrelatedness of agents and commitment to values beyond rules institutionalized by the state.

The SERRC's collective vision may well remain unrealized given both widely varied and continually added statements from its members. Still, as the SERRC is inspired by Fuller's social epistemology, I see the SERRC's future as intertwined with Fuller's current and future work, which concerns the interplay of scientific knowledge and changing conceptions of humanity.

NOTE

1. blogs.lse.ac.uk/lsereviewofbooks/2013/05/07/book-review-humanity-2-0/.

BIBLIOGRAPHY

Barron, Colin, ed. "A Strong Distinction Between Humans and Non-Humans is no Longer Required for Research Purposes: A Debate between Bruno Latour and Steve Fuller." *History of the Human Sciences* 16, no. 2 (2003): 77–99.

Basbøll, Thomas. "I, Social Epistemologist." *Social Epistemology Review and Reply Collective* 3, no. 4 (2014): 14–15. wp.me/P1Bfg0-Y9.

Collin, Finn. "Two Kinds of Social Epistemology." *Social Epistemology Review and Reply Collective* 2, no. 8 (2013): 79–104. wp.me/p1Bfg0-Ul.

D'Agostino, Fred. "How Can We Collectivize a Set of Visions about Social Epistemology?" *Social Epistemology Review and Reply Collective* 3, no. 8 (2014): 5–9. wp.me/P1Bfg0-Y9.

Fuller, Steve. *The Governance of Science: Ideology and the Future of the Open Society*. Buckingham, UK: Open University Press, 2000.

———. *The Intellectual*. Cambridge: Icon Books, 2005.

———. *The New Sociological Imagination*. London: Sage Publications, 2006.

———. *New Frontiers in Science and Technology Studies*. Cambridge, UK: Polity Press, 2007.

———. *Dissent over Descent*. Cambridge: Icon Books, 2008.

———. "Humanity: The Always Already—Or Never To Be—Object of the Social Sciences." In *The Social Sciences and Democracy*, edited by Jeroen Van Bouwel, 240–64. Basingstoke: Palgrave Macmillan, 2009a.

———. "In Search of the Sociological Foundations of Humanity." *History of Human Sciences* 22, no. 2 (2009b): 138–45.

———. *Humanity 2.0. What it means to be Human Past, Present and Future*. New York and Basingstoke: Palgrave Macmillan, 2011.

———. *Knowledge: The Philosophical Quest in History*. New York: Routledge, 2014a.

———. "Social Epistemology: The Future of an Unfulfilled Promise." *Social Epistemology Review and Reply Collective* 3, no. 7 (2014b): 29–37. wp.me/p1Bfg0-1wG.

———. "Towards a Proactionary Welfare State." *Social Epistemology Review and Reply Collective* 3, no. 5 (2014c): 82–84. wp.me/p1Bfg0-1sx.

Kerr, Eric. "Are You Thinking What We're Thinking? Group Knowledge Attributions and Collective Visions." *Social Epistemology Review and Reply Collective* 3, no. 1 (2013): 5–13. wp.me/P1Bfg0-Y9.

Martini, Carlo. "The Problem of Disagreement and Social Epistemology." *Social Epistemology Review and Reply Collective* 3, no. 8 (2014): 22–24. wp.me /P1Bfg0-Y9.

Pedersen, David Budtz. "Who Should Govern the Welfare State 2.0? A Comment on Fuller." *Social Epistemology Review and Reply Collective* 2, no. 12 (2013): 51–59. wp.me/p1Bfg0-19F.

Remedios, Francis. *Legitimizing Scientific Knowledge.* Lanham, MD: Lexington Books, 2003.

———. "Fuller and Mirowski on the Commercialization of Scientific Knowledge." In *The Social Sciences and Democracy*, edited by Jeroen Van Bouwel, 229–39. Basingstoke: Palgrave Macmillan, 2009.

———. "Review of *Humanity 2.0. What it means to be Human Past, Present and Future.*" *LSE Review of Books.* May 2013. blogs.lse.ac.uk/lsereviewofbooks/2013/05/07/book-review-humanity-2-0/.

Chapter 4

A Social Epistemology for Scientific Excellence

David Budtz Pedersen

In recent years the notion of scientific excellence, and its interconnectedness with the capacity to perform basic research at the highest international level, has become exceptionally influential in science policy. New instruments for competitive funding as well as new funding bodies have been established to stimulate long-term and high-risk research. Research excellence initiatives have become a major driver for achieving breakthroughs in basic research and scientific and technological developments through the provision of large-scale, long-term funding, and state-of-the-art research infrastructures. This chapter envisages a new role for the social epistemologist in shoring up priorities and recommendations for balancing scientific excellence with broader societal needs.

BALANCE BETWEEN EXCELLENCE AND RELEVANCE

There is widespread consensus that a central goal of the research community is to serve the ends of society, helping to construct more effective policies, and facilitate public and private innovation. However, there is disagreement not only with regard to the means to reach those goals but also with regard to the methods and indeed the balance between long-term and short-term goals. Only few scholars would disagree that university research ought to be relevant to societal needs, such as those in a welfare state. The question is how, whether by allowing researchers to follow their own curiosity (in the classical autonomy mode of research) or whether organizational, administrative, and budgetary management is needed to ensure greater societal relevance.

This discussion is often cast in terms of reaching a balance between basic research (or scientific excellence) and societal impact.

For historical reasons, basic research has been under pressure in the past decades. Since World War II, science policies in countries of the Organisation for Economic Co-operation and Development (OECD) have increasingly focused on strategies and instruments that aim at stimulating scientific innovation. Innovation policy has become an intrinsic part of science policy and is today built into the regulatory system that guides decisions concerning funding, evaluation, and commercialization of research. With the advent of "the knowledge society" (Stehr 1994) and "the knowledge-based economy" (Foray and Lundvall 1996) governments and research funding agencies have sought to develop coherent measures for realizing the social and economic benefits of science (Martin and Tang 2007). Simultaneously, policy makers and science administrators have emphasized the strategic importance of evaluating research and innovation systems in terms of their contribution to economic performance and other output factors that are perceived as essential in the global competition for knowledge.

Two historical examples illustrate this shift in emphasis from basic research to applied and mission-oriented research: on the one hand, the abandonment of the linear model of innovation, and on the other hand the emergence of biotechnology and bioscience as the scientific ideal after 1989. In the United States, Vannevar Bush's 1945 report "Science: The Endless Frontier" had a specific status in defining an agenda for postwar science policy. Bush believed in the trickle-down effects that emerge when basic research translates into innovation. He believed the role of the state was to correct market failures by investing directly in public goods such as basic research that later would spill over "to our security as a nation, to our better health, to more jobs, to higher standard of living, and to our cultural progress" (Bush 1945).

The linear model has since supported numerous criticisms concerning the linearity of the innovation process. Various metaphors and policy concepts such as a "Mode-2 knowledge-production" (Gibbons et al. 1994), the "risk society" (Beck 1992), "the triple helix of university-industry-government relations" (Etzkowitz and Leydesdorff 2000), "post-academic research" (Ziman 2000), and "innovation ecosystems" (Bowonder and Miyake 2000) have provided alternative (nonlinear) theories of innovation. Common among these approaches is the observation that societal impact and application-oriented research are not things that emerge at the end of the research process, but are built into the research design from the beginning. Studies of pharmaceuticals and biotechnology companies in the 1990s prompted a number of new and more contextualized theories of knowledge diffusion and technology transfer. These developments made the distance from basic research to commercial application much shorter. They depicted a new and emerging science system in which short-term societal and economic needs trump the classical doctrine of curiosity-driven, "blue-sky" research (Lundvall and Borrás 2005).

The advent of Mode 2 or post-academic research indicates that institutional boundaries are becoming blurred, and that researchers are engaging in new collaborative configurations that are formed around external problems as their epistemological and pragmatic guide, working on their solution in specific "contexts of application" (Nowotny 2006). In such research programs, funding is allocated according to a set of specific objectives, for instances, solving "grand challenges" or delivering key enabling technologies, such as nanotechnology, biotechnology, and information technology. Methods for defining strategic priorities include scenario research, foresight and forecasting, anticipatory governance, and participatory priority setting (Guston and Sarewitz 2002). Because of the need to legitimize public research spending, such policy tools have become valuable instruments when ensuring political accountability. They follow a larger trend within public administration that emphasizes multilevel governance and New Public Management (Hood 1991; Powell and DiMaggio 1991).

RENAISSANCE OF BASIC RESEARCH

While the top-down model of research priority setting has become increasingly popular, rising needs for basic research and capacity building have at the same time underscored the necessity of providing long-term "bottom-up" funding. The organized production and use of knowledge for the purpose of industrial innovation is an important subdynamic of the current socio-economic system. But maintaining a leading position in a highly competitive knowledge economy requires a high level of scientific proficiency and excellence in basic research. From a period in which emphasis was primarily given to the application of knowledge in external contexts, basic research has increasingly seen a renaissance in science policy.

New instruments and funding bodies (such as the Agence Nationale de la Recherche in France and the European Research Council [ERC]) have been established in order to stimulate risk taking and achieve breakthroughs in basic research (Celis and Gago 2014). Under "Horizon 2020," the current program of EU funding for research and innovation (2014–2020), the ERC has seen a significantly increased budget of more than €13 billion. This is an explicit recognition by the European Union of the pioneering role that the ERC plays in Europe, promoting scientific excellence and contributing to capacity building in numerous European institutions.

At the international level, several countries have recently launched new programs in order to increase the excellence of their respective research and higher education institutions. For instance, in Germany, the "Initiative on Excellence" was commenced in 2007. It aims at strengthening Germany as a science and research nation, improving Germany's international competitiveness, and making cutting edge research at German universities and laboratories more visible. In France, the French National Research Agency (ANR) has implemented a major program called "Investments for the Future." The main focus is on ten-year projects opening up new perspectives and nurturing collaborative projects that would otherwise never have been realized. In Denmark, the Danish National Research Foundation (DNRF) has supported the long-term promotion of outstanding researchers under a high-trust regime primarily through its main funding instrument: the Centre of Excellence scheme (Krull et al. 2014).

Yet, there exist considerable differences with regard to the technical definitions of scientific excellence, and even more so with regard to the means and arrangements needed to foster excellence in basic research. A recent OECD survey of global research excellence initiatives found fifty-six different funding schemes across eighteen countries. Among the different schemes, four principles for stimulating excellence in science were identified: (i) create an environment for improved quality of research; (ii) increase the international visibility of national research centers and universities; (iii) recruit the best international talents and early-career researchers; and (iv) support resource-intensive research and capacity building, including interdisciplinary research and collaboration (OECD 2014).

In a certain way, these principles are trivial. The fact that international visibility and reputation are decisive for attracting and maintaining talents, or that provision of long-term funding and research autonomy are crucial for scientific and scholarly excellence, corresponds to a long-standing ideal of autonomy in science that does not leave much room for active policy making. As Helga Nowotny perceptively remarks in her reading of John Ziman's *Real Science*, basic research may not even by a policy category. Basic research is often defined by exclusion, Ziman argues, a residual category of activities that do not fall under instrumental, or strategic purposes. In the end, the notion of basic (or excellent) science is difficult to define in policy term because policy is often about active involvement, management, and practical intentionality, whereas scientific excellence is about long-term, blue-sky, and self-organizing social practice (Nowotny 2006).

THE ROLE OF THE SOCIAL EPISTEMOLOGIST

Following this line of reasoning it is not surprising that science and technology studies (STS) and science policy studies (SPS) for the last decades have focused almost exclusively on policy-relevant research—and only rarely on the conditions required for stimulating progress in basic science. Sociologists of knowledge have devoted considerable attention to the governance of science in relation to different pragmatic, industrial, and policy-oriented goals, whereas studies of conceptual innovation, theory-driven science, and curiosity-oriented research largely have been neglected (Rip and van der Meulen 1997; Jasanoff 1997; Cozzens et al. 1990; Gibbons et al. 1994; Jasanoff 2002; for an exception, see Pickering 1995). Here, I want to suggest a new central role for social epistemology. Social epistemology is often portrayed as the normative analysis of burdens and benefits of scientific knowledge production. But social epistemology also entails the ability to study scientific excellence and cognitive competences with regard to researchers, research infrastructures, and research institutions; to emphasize the coordination of research programs and priorities and knowledge sharing between public and societal actors; and to examine the role of research institutions and research management practices, putting emphasis on the role of social psychology and organizational studies in organizing research teams and opening research up to interdisciplinary collaboration and cross-fertilization (Fuller 2000; Fuller 2002; Langfeldt 2001; Lamont 2009; Holbrook and Hrotic 2013; Holbrook and Frodeman 2011).

Here is a first attempt at formulating a research agenda for a social epistemology of scientific excellence. The founding principle of scientific excellence is curiosity-driven research combined with individual and collective scientific creativity as its driving force. Krull et al. (2013) suggest that one of the key conditions for creating excellent scientific teams is to provide them with sufficient funds, a long-term funding perspective, and grant them far-reaching autonomy with respect to the research agenda. This formula should further be broken down into a list of elementary socio-epistemic conditions that needs to be in place in order to stimulate innovative and successful research:

1. *Competence*: Empirical research is needed to examine how researchers develop their skills in a prolific learning environment, and which practices and strategies are available in order to enhance the cognitive division of labor within a research team. This is a basic prerequisite without which no further scientific development appears to be possible.
2. *Creativity*: Achieving excellence in science requires creative thinking. Often only considered as a "collateral" aspect, scientific and social creativity is important components in organizing research teams. Social studies of scientific creativity should be developed with the purpose of finding the best conditions to enable creative and unconventional research projects.
3. *Commitment*: Not only practicing researchers, but also institutional managers and funding agencies must make a real commitment in order to

achieve high-impact research. More knowledge is needed about the optimal fit between management, funding, and research practice. Empirical studies have already documented perverse effects when researchers are subjected to rigid incentive structures. Institutional commitment and financial security should be provided to encourage scientists to enter new fields and leave common wisdom.

4. *Communication*: Developing an epistemology of scientific excellence should include an examination of the role of conversation and communication in achieving scientific advancement, in particular in interdisciplinary and transcultural settings. Such communication processes are often not seen as the central aspect of research projects, but their inspirational and epistemic value should not be underestimated.
5. *Collaboration*: Working closely with other scientists and partners often leaves a positive impact on scientific projects. Research in social epistemology can help produce more coherent models and measures of transnational and interdisciplinary collaboration, including models that encourage researchers to engage nonacademic actors and institutions that will benefit from getting access to new and innovative research results.
6. *Continuity and organizational stability*: Forging new paths in a barely known territory often take longer than only two or three years (the usual running time of research projects). Empirical and conceptual studies are needed to understand the role of mistakes, nonresults, and negative results in basic research, and provide models of social practices that allow for changes in direction and theory choice.

Besides this generic research agenda for a social epistemology of scientific excellence, it is necessary to highlight the *unintended* side effects of shifting research priorities toward more competitive modes of research funding. There is a risk that international competition is eroding the national capacity of research institutions to deliver excellence in all disciplinary breadth. There is no substantial commitment to ensure favorable and steady career paths for young researchers or to facilitate long-term support structures for early-career researchers that allow genuine mobility. Fostering competition and structural change also create friction and other unintended consequences. Competitive funding means that some fields (i.e., the humanities and social sciences) may be disproportionately disadvantaged while others may experience an overflow of support and investment. There is a danger of concentrating resources excessively on too few institutions or research topics without securing epistemic diversity and complementarity with other institutional goals. Obstacles that may undermine drivers for excellence include international bandwagon effects (science bubbles); simplistic benchmarking exercises that may counteract renewal and creativity; too short-term perspectives on return on investments; and the enforcement of hierarchical structures between researchers that may stifle innovation and the proliferation of disruptive ideas (Pedersen and Hendricks 2013).

Despite the rising interest in promoting scientific excellence through special instruments and funding models, these critical issues need further clarification and attention. Social epistemologists can offer a naturalistic account of what scientific excellence is. Among other things, this requires the epistemologist to introduce a sociological dimension, not to replace the traditional philosophical dimension, but to enlarge it. The organization and governance of science entails sociocultural elements as well as cognitive elements. A study of scientific excellence can also form part of a broader inquiry into the ways in which sources of funding and other research groups are affecting scientific performance. This type of policy-oriented social epistemology can investigate such questions as: How do policies square with researchers' needs, and what are the implications for policy? Do researchers adhere to externally set standards of quality and accountability in their research? How do they perceive the impact of funding sources and reward systems on their research performance (Morris 2000)? This line of epistemological research includes an assessment of internally generated pressures—for example, from conceptual developments in science—as well as external pressures stemming from incentive structures or research management.

In short, social epistemologists should be concerned with understanding and constructing research excellence initiatives while at the same time contributing to a more coherent framework for promoting basic scientific knowledge. The renewed focus on research excellence, as well as the extensive funding available for excellent scientists worldwide, promises a shift away from short-term strategic research. Instead a more modest, context-sensitive, long-term epistemology of science policy is evolving—taking as its point of departure the actual research practices of frontier scientists.

BIBLIOGRAPHY

Beck, Ulrich. *Risk Society: Towards a New Modernity*. London: Sage Publications, 1992.

Bowonder, B. and T. Miyake. "Technology Management: A Knowledge Ecology Perspective." *International Journal of Technology Management* 19, nos 7–8 (2000): 662–84.

Bush, Vannevar. *Science, the Endless Frontier: A Report to the President*. Washington, DC: U.S. Government Office, 1945.

Celis, Julio E. and Mariano Gago. "Shaping Science Policy in Europe." *Molecular Oncology* 8 (2014): 447–57. doi:10.1016/j.molonc.2014.03.013.

Cozzens, Susan E., Peter Healey, Arie Rip, and John Ziman, eds. *The Research System in Transition.* Dordrecht: Kluwer Academic, 1990.

Etzkowitz, Henry and Loet Leydesdorff. "The Dynamics of Innovation: From National Systems and 'Mode 2' to a Triple Helix of University Industry-Government Relations." *Research Policy* 29, no. 2 (2000): 109–23.

Foray, Dominique and Bengt-Ake Lundvall. "The Knowledge-Based Economy: From the Economics of Knowledge to the Learning Economy." In *Employment and Growth in the Knowledge-Based Economy*, edited by the Organisation for Economic Co-Operation and Development, 11–32. Paris: OECD, 1996.

Fuller, Steve. *The Governance of Science*. London: Open University Press, 2000.

Fuller, Steve. *Knowledge Management Foundations*. Boston, MA: Butterworth-Heinemann, 2002.

Gibbons, Michael, Camille Limoges, Helga Nowotny, Simon Schwartzman, Peter Scott, and Martin Trow. *The New Production of Knowledge: The Dynamics of Science and Research in Contemporary Societies*. London: Sage, 1994.

Guston, David H. and Daniel Sarewitz. "Real-time Technology Assessment." *Technology in Society* 24, nos. 1–2 (2002): 93–109.

Holbrook, J. Britt and Robert Frodeman. "Peer Review and the *Ex Ante* Assessment of Societal Impacts." *Research Evaluation* 20, no. 3 (2011): 239–46. doi:10.3152/095820211X12941371876788.

Holbrook, J. Britt and Steven Hrotic. "Blue Skies, Impacts, and Peer Review." *ROARS Transactions: A Journal of Research, Policy, and Evaluation* (July 21, 2013). doi:10.13130/2282-5398/2914.

Hood, Christopher. "A Public Management for All Seasons?" *Public Administration* 69 (1991): 3–19.

Jasanoff, Sheila, ed. *Comparative Science and Technology Policy*. Cheltenham: Edward Elgar, 1997.

Jasanoff, Sheila, ed. *Handbook of Science and Technology Studies.* 2nd ed. London: Sage, 2002.

Krull, Wilhelm et al. *Evaluation of The Danish National Research Foundation*. Copenhagen: Danish Ministry of Science, Innovation and Higher Education, 2013.

Lamont, Michèle. *How Professors Think: Inside the Curious World of Academic Judgment*. Cambridge, MA: Harvard University Press, 2009.

Langfeldt, Liv. "The Decision-Making Constraints and Processes of Grant Peer Review, and Their Effects on the Review Outcome." *Social Studies of Science* 31 (2001): 820–41.

Lundvall, Bengt-Åke and Susana Borrás. "Science, Technology and Innovation Policy." In *Innovation Handbook*, edited by Jan Fagerberg, David C. Mowery, and Richard R. Nelson, 599–631. Oxford: Oxford University Press, 2005.

Martin, Benn R. and Puay Tang. *The Benefits from Publicly Funded Research*. SPRU Working Paper No. 161. University of Sussex, 2007.

Morris, Norma. "Science Policy in Action: Policy and the Researcher." *Minerva* 38, no. 4 (2000): 425–51.

Nowotny, Helga. "Real Science is Excellent Science—How to Interpret Post-Academic Science, Mode 2 and the ERC." *Journal of Science Communication* 5, no. 4 (2006): 1–3.

OECD, Organisation for Economic Co-operation and Development. *Promoting Research Excellence: New Approaches to Funding*. Working Party on Research Institutions and Human Resources. DRAFT October 14, 2013. Paris: OECD.

Pedersen, David Budtz and Vincent F. Hendricks. "Science Bubbles." *Philosophy and Technology*. Springer (online-first). 2013. doi:10.1007/s13347-013-0142-7.

Pickering, Andy. *The Mangle of Practice: Time Agency and Science*. Chicago: University of Chicago Press, 1995.

Powell, Walter W. and Paul J. DiMaggio, eds. *The New Institutionalism in Organizational Analysis*. Chicago, IL: University of Chicago Press, 1991.

Rip, Arie and Barend van der Meulen. "The Post-Modern Research System." In *Science in Tomorrow's Europe*, edited by Remi Barré, Michael Gibbons, Sir John Maddox, Ben Martin, and Pierre Papon, 51–67. Paris: Economica International, 1997.

Stehr, Nico. *Knowledge Societies*. London: Sage Publications, 1994.

Ziman, John. *Real Science: What It Is and What It Means*. Cambridge: Cambridge University Press, 2000.

Chapter 5

From Social Epistemology to Reflexive Sociology

Inanna Hamati-Ataya and Stephen Norrie

The demand for reflexivity in social science is a demand for self-knowledge, for a sociology of sociology. To what ends? Alvin Gouldner, along with Pierre Bourdieu, the great prophet of "reflexive sociology," used to breathe thunder at anyone suggesting that it amounted to mere "navel gazing." But if what is imagined is an endless stream of ethnographies of seminars and faculty meetings, of discourse analyses of discourse analyses, justified only by the fact that these too are "social," what else could one think? Do sociologists not have a duty to confront the real social problems—to go out and see the blood in the streets?

Sociologists may seem irrelevant here because they lack accumulated social power of the kind wielded by capitalists, or union leaders, or emerging social movements—but might they not aspire to an intellectual power? If the aspiration to a social science means anything, it must mean the possibility of society acquiring an objective knowledge of itself, or rather, of its members acquiring an enlightened understanding of their own social situations and interdependencies. The question of the existence or nonexistence of such a possibility, however, has wide-reaching consequences for the interpretation of particular social facts. It makes a great deal of difference to the understanding of any social problem or conflict, and of available remedies, if those who participate in and suffer from it are, or are not, deemed *capable* of correctly understanding their situation. Such understanding does not come by nature: society is a complex whole, and the individual's situation is determined by the total social process. Self-knowledge thus requires the ability to draw on specialist knowledges, and the existing social sciences (academic or not) thus represent its best available tools, albeit still imperfect ones. If sociologists exclude a systematic reflection on their own social role and practices, that may systematically distort their understanding of the world they claim as their object domain. For this reason, reflexive sociology is a precondition of pursuing a coherent and ethical social science.

In trying to understand what this might mean in practice, the slogan of reflexivity may confuse more than it clarifies: taken alone, it is so abstract that it is capable of applying to virtually anything, and this makes it a handy device for assimilating projects that have little else in common. Since Kant, at least, philosophers have recognized that self-consciousness is an ineliminable constituent of any objective consciousness whatsoever, and his successors extended the insight to all organic processes, insights that are now staples of modern psychology and biology. In sociology, reflexivity can indicate a generic character of human being, the image management prevalent in late modern cultures, the standard control procedures of modern firms and states, and even an avant-gardist stylistic device. But even if relatable, these uses hardly add up to a single topic. At times, moreover, the demand for "reflexivity" degenerates to a ritual confession of perceived prejudices (but what of the unperceived ones, i.e., the real prejudices?), which has little effect on the work itself.

It is necessary, therefore, to stipulate with greater methodological precision what a reflexive orientation entails. In particular, it is necessary to distinguish sociological from philosophical reflexivity, at least insofar as that is defined by the classical epistemological tradition, which centered on *highly abstract* reflection on a desocialized and dehistoricized "mind in general"—an idealized universal "subject." The philosophy of science, with its concern to abstract the "logic of science" from the concrete social institutions to which it was connected, in the end represented a fairly superficial break with this tradition. The abstracted "minds" or "logics" are, in both cases, merely naturalized expressions of particular social arrangements, and serve to obstruct a critical social understanding of science, and the development of a responsible, and accountable social science suitable for an enlightening social mission.

There are several equally appealing theoretical alternatives to ground the project of a reflexive sociology. In this chapter, Inanna Hamati-Ataya first argues that until some fundamental obstacles and resistances to sociology itself are defeated, reflexivity has very little hope of shaping our understanding of ourselves and of the world. The first step of a reflexive sociological project should therefore be to examine the conditions of possibility of reflexive sociology itself. Stephen Norrie then explores how critical theory, which explicitly advocated the organization of knowledge around the needs and capacities of the present rather than by reference to idealized knowers or bodies of knowledge, never managed to realize this intention, largely because of its omission of a developed sociological self-understanding. He attempts to demonstrate one way in which a reflection on these practices can be transformative for the scientific understanding of society as a whole, through advocacy of a more collective approach to synthesis work in the social sciences—a theme that returns us to the purposes of the Social Epistemology Review and Reply Collective (SERRC) project itself.

BEFORE REFLEXIVITY: ON THE LIMITED EFFICACY OF "THE SCANDAL," INANNA HAMATI-ATAYA

The consensus over social epistemology's meta-epistemological and normative rationale makes it possible for the members of the SERRC to formulate and pursue a "collective vision," of which this volume is the first organized expression. Regardless of how each of us defines their own individual socio-cognitive or thematic interests, this consensus makes it unnecessary for our internal conversations to address the "basics" of our shared assumptions. And yet this volume gives us the opportunity to engage in an exercise of self-clarification aimed at a new audience—and a potentially sympathetic one, too. More importantly, it *is* necessary for any group, while it purposefully pursues its forward-looking agenda, to constantly keep an eye on those structural obstacles that systematically defeat it. These obstacles become salient once one steps out of the community of social epistemologists, in any of the subfields that populate the humanities and social sciences wherein knowledge constitutes an object of cognitive and/or normative interest, or that calls on classical epistemology to validate/legitimate its knowledge claims and arbitrate among its competing theoretical orientations. Specifically, the case for reflexivity makes obvious those mental attitudes that reject the notion that knowledge can—let alone should—be studied like any other ordinary object of human life, and/or that knowledge-related normative discourses and actions can—let alone should—be grounded in just such a realist, empirical, sociological investigation of knowledge.

It is, then, in relation to the *sociology of knowledge* as one of the two components of social epistemology that the obstacles to social epistemology need to be defined first and foremost. More importantly, if reflexivity pertains to social science's ability to illuminate the social conditions, situations, and interdependencies that make it possible and underscore its representations and actions, then reflexivity dictates that science should interrogate the conditions of its *im*possibility as well. Social epistemologists and sociologists of knowledge therefore have more than a mere political or strategic interest in identifying what defeats or undermines their purposes: they have a *logically necessary cognitive interest* in understanding the objective factors that enable or on the contrary impede social-epistemological/sociological thinking. In fact, this section is informed by an epistemological tradition that predates social epistemology and that takes obstacles at precisely the starting point of a systematic reflection on the problem of knowledge. It is what Lecourt (1972), following Canguilhem, has called the French tradition of "historical epistemology," established by Bachelard and pursued by Canguilhem and Foucault, but also providing the main inspiration for one of the most systematic and explicit formulations of reflexivity in our time, namely, Bourdieu's:

> Understood as the effort whereby social science, taking itself for its object, uses its own weapons to understand and check itself . . . [Reflexivity] is not a matter of pursuing a new form of absolute knowledge, but of exercising a specific form of epistemological vigilance, the very form that this vigilance must take in an area *where the epistemological obstacles are first and foremost social obstacles.* (Bourdieu 2004, 89; emphasis added)

While Western philosophy and epistemology have no doubt been preoccupied with many different kinds of obstacles to knowledge and knowing, ranging from physiological/perceptual to ideological ones, the focus on the *technical* dimension of the processes whereby these obstacles impede knowledge has systematically obscured their *social and political* dimensions. And it is precisely because of their ability to "problematize" the question of knowledge in sociopolitical terms that scholars like Canguilhem, Foucault, and Bourdieu—and before them Hobbes, Marx and Engels, Mannheim—have managed to transform reflexivity from an analytical-logical and formal feature of philosophical self-understanding, to a truly sociological understanding of the socially constituted collective knowing subject. Moreover, as Bourdieu repeatedly pointed out, nowhere are sociopolitical obstacles to knowledge more perniciously and efficiently hidden than when they are presented as mere technical, philosophical ones. We should, then, attempt to trace these obstacles to sociological reflexivity from their superficial philosophical appearances to their deeper sociopolitical dimensions and functions. The purpose of such a reflexive investigation should be to determine how to efficiently get past some psychological, intellectual, or dispositional resistances to the mindset and project of reflexivity. This entails understanding the structural factors that systematically defeat the possibility and progress of sociological and empirically based normative thinking about thought, knowledge, and science.

The obvious apparent characteristics of the mode of thinking that undermines sociological reflexivity are the idealist, a priori, analytical, and logical-deductive features that characterize the philosophy that dominates the Anglo-American world today—and by extension the areas subjected to its epistemic violence. Because it shapes the dominant schools of epistemology and philosophy of science that are imported into the humanities and social sciences to arbitrate and regulate their internal metatheoretical discourse, it effectively pollutes most of the disciplinary landscape, making it impermeable to the very notion, and overall project, of a sociology of knowledge (and more broadly, to a sociology of thought, ideas, and cognition). Where its pernicious effects have been revealed through sociohistorical analysis, it is resisted by those who, armed with a materialist or realist approach to social reality, have in effect understood that social scientists needed to become their own epistemologists, simply because knowledge and science can neither be understood in the singular and with a capital letter, nor be fantasized about as if they were outside of history and society, nor be abstracted as mere linguistically embodied systems of ideational representations producing ruled behavior. By stubbornly ignoring the fact that knowledge*s* and science*s* are the real signifieds that need to be defined and understood; that they are historically situated and socially produced *practices* that create norms rather than merely the rationalized products of aprioristically established ones; and that their social authority and authoritativeness are precisely dependent on their social nature being (rendered) invisible, idealism in effect serves as the mediator, legitimator, and enforcer of the social order that reflexivity is meant to unmask to our consciousness.

The dominant idealist view is a serious obstacle to the development of the sociology of knowledge—and by extension, social epistemology—as a foundation for an alternative (reflexive and empowering) way of thinking about and practicing the social sciences, *and* becoming a socially useful scholar. And in the face of idealism, one's interpretivist and constructionist sensibilities often require additional support from older, more positivist traditions—those same traditions that allowed thought, in a properly social-*critical* manner, to oppose commonsensical, traditional, and religious narratives about society, thereby bringing society within the realm of conscious, autonomous thought. And these positivist reflexes should not be restrained whenever someone laughs at the mention of such obscure things as the sociology of logic and mathematics, or the social conditions of possibility of a given philosophical "-ism." Unlike Foucault's (2001) laughter, which expressed an uncomfortable awakening to what lied beyond epistemic consciousness, these idealists' laughter is that of confident ignorance: another manifestation of the fact that neither knowledge, nor thought, have yet been seriously and systematically addressed as "social phenomena" endowed with real ontological status that *could*—let alone *should*, as per Durkheim's (2014) rule—be "treated as things."

While philosophers might not feel compelled to treat anything as a social phenomenon, theorists who work within the social sciences, and who actively contribute to the philosophy of social science as it pertains to their field or draw on it for their research, might be expected to extend sociological thinking to their own conceptual and methodological tools. And yet the dominance of idealism within the social sciences creates a disjunction between social scientists/theorists' acceptance of knowledge as a social phenomenon in ontological terms, and their implicit or explicit rejection of its social status at the epistemological level of inquiry. One obvious indicator of this disjunction is the extent to which theoretical or empirical studies about the relation of knowledge to social interests, order, and power actually impact the epistemological discussions of those fields where such studies have gained some measure of authority. Even when feminist (standpoint), constructivist, Marxist, poststructuralist, critical (Frankfurt-School-style), or postcolonial approaches are considered relevant and informative with respect to how given social structures and processes enable given cognitive configurations and systems, there is always a limit to how such relevance is brought to bear on metatheoretical and epistemic discussions. So one can very well accept, say, Foucault's analysis and conclusions on the constitution of psychiatry as a cognitive field, and use this analysis to illuminate other social aspects of "power-knowledge" at work, but the logical step of extending such analysis to the philosophy of science and epistemology is difficult to initiate and pursue. The sociological "trickles up" with great difficulty, as if these fields were outside of social reality and history. And without this necessary step, what does a "critical" or "reflexive" scholarship mean at all?

For some mysterious reason, the assumption remains that discussions of metatheory should themselves be theory driven or theoretical (and hence self-contained and self-sustained) rather than empirically grounded, as if idealist regress in analysis were a conceptual or methodological necessity (it certainly appears to be a *logical* necessity for epistemologists, and so idealism begets idealism, and the analytical reinforces the analytical). The typical reaction on the classical side—as noted by Steve Fuller (1988) early on—is still very much alive: the sociology of knowledge/science cannot tell us anything useful about epistemic or epistemological questions; it does not provide us with a "demarcation" criterion between knowledge/science and opinion/nonscience, nor does/can it tell us what knowledge/science "are." This posture still permeates dominant metatheoretical debates in the social sciences—especially in political science. When theorists, for instance, gather to discuss what a theory is, and how to produce "better" theories, they take their cues from the classical debates and categories of the philosophy of science (and end up reaching the same dead ends), most of which start from some a priori notion, principle, relation, or standard of validity, measure, and inference. It is never clear whether and how these models informed the actual development of *real* and "good" social theories, or why one would not rather turn to the *actual* history of sociological explanations in order to critically induce the criteria or conditions that make a social theory a "good" or useful one (which is what a Bachelardian posture would entail if systematically applied to all the constituents of scientific practice).

Such questions are implicitly rejected on the basis that what *is* cannot inform what *ought to be*. Indeed when classical philosophers/epistemologists, and those who follow their lead, ask what knowledge "is," they really are asking a different, that is, *normative*, question. This same idealism, whereby reality (both natural and social) is de facto considered irrelevant in answering questions about the "nature" of things and processes, shapes the way political and normative theories are understood, taught, and used in the Anglo-American tradition—for example, Rawls's theory of justice, or that peculiar field called "*applied* ethics," where a view of "the good" and "the just" is constructed independently of any engagement with the history and sociology of norms, and then forced onto political reality to make it fit the model. It is unlikely that if epistemologists or political and normative philosophers in that tradition were asked what DNA "is," they would rely on their mode of reasoning to answer such a question. But somehow, when it comes to knowledge, science, theory, objectivity, values, and norms, the relation between truth and reality, thought and practice, is reversed.

This attitude is especially problematic in political studies, where "power" and "the political" are considered as core objects. A nonsocial and hence depoliticized understanding of thought, knowledge, and science at the epistemological level of inquiry is here not only logically incoherent, but also practically problematic and dangerous. It is interesting that a discipline that has dealt for so long with such notions as "ideology" and the "knowledge-power nexus" has managed to retain so much idealism and a priori thinking when it comes to epistemic and ethical questions. Some studies on the socio-institutional development of political thought in specific locations provide important elements to answer this question, but more needs to be done to unravel the general patterns that enable the persistence of idealism. In the meantime, one need look no farther than the foundational texts in sociology and the sociology of knowledge to get a sense of what is at stake. Especially illuminating is the fact that their authors' posture and clarifications on the one hand, and the objections and outrage their intellectual moves triggered on the other, still resonate with the contemporary situation. So when going back to Comte, Marx, Durkheim, Bachelard, as well as Weber, Scheler, and Mannheim, and reading their texts not merely as cognitive "representations" but fundamentally as social "interventions," our age does not feel so different from theirs: the sociology of knowledge is still a "scandal"—even when it is only partially understood, and hence only intuitively perceived as a threat or disturbance to dominant ways of thinking and doing. But more generally, it seems that sociology itself is still scandalous, and its potentially socially subversive nature still very strong.

Pedagogical practice can be especially informative as to the extent to which sociological thinking has really impacted contemporary thought, as opposed to the merely policy-related appeal of the social sciences and their methodologies. Even when students develop affinities with constructivist or poststructuralist approaches to social reality, their cognitive reflexes remain somehow "below" sociological thinking. Perhaps the way poststructuralism and postmodernism have developed into intellectual and university fashions outside of their socio-intellectual context of emergence is partly responsible for this. Perhaps a more important factor is the ways wherein sociology informs secondary education, especially in comparison to philosophy, and hence whether sociology's "unmasking" effect is able to efficiently operate at the formative level where total-ideology-based "commonsense" is institutionally reinforced. Whatever the case may be, it is quite perplexing how students who can handle de/constructionist or genealogical thinking at quite a sophisticated level are puzzled by such a basic notion that suicide, aesthetic preferences, love, or mathematics can be studied/considered as social phenomena. Indeed, this makes it difficult to decide whether to offer students the most up-to-date research on issues pertaining to knowledge and power, or just to send them straight (back) to Comte's *Lectures* and Durkheim's *Rules*.

A more troubling issue is how resistance to sociological thinking seems to be related organically to liberalism and its associated individualism and universalism. One should be cautious about claiming a necessary link between sociology and "the left," even if the birth of (European) sociology is undoubtedly connected to socialism and a more or less organic view of society (as opposed to American sociology, and early British social theorizing). But the dominant liberal ideology has certainly shaped everyday thinking about social and moral acts and facts, and what our "commonsense" relation to social reality has become. And it certainly supports, and is reinforced by, analytical-idealist thought, as they both share the ability to sustain a mysterious connection between the unverifiable private realm of the individual (and her intimate thought, values, and personal preferences) and the transcendental realm of the universal (thought, values, and norms in "the absolute")—both of which are characteristically defined as a-social, and hence as beyond the purview of sociological thought and inquiry.

Are we, then, to eternally rediscover Marx and Engels's demonstration of the ideological nature of idealism without any hope of defeating idealism not simply as the mediator of the existing order, but also as an obstacle to reflexive thought? If thought is shaped by structures and practices then surely it is through a transformation of structures and practices that one should perform one's reflexivity. And this is what reflexive sociology enables us to do.

CAN CRITICAL THEORY BE COMPUTERIZED AND COLLECTIVIZED? STEPHEN NORRIE

The idea of a "critical theory," an emancipatory knowledge distinct from academic "traditional theory," can probably best be explained in its context of origin. The idea expresses the feeling of young central European intellectuals who had been radicalized by the revolutionary movements of the early twentieth century, but who had come to feel, with their failure in the early 1920s, that their liberatory promise would not, and perhaps could not, be realized.[1] Identification of a specifically emancipatory type of theory thus served both to outline an anti-Stalinist research program, and also to express the novel state of alienation in which these young intellectuals found themselves: the revolution had given them a sense of a new liberatory form of knowledge, directly linked to practice, but alienated from the workers' movement, they now had no such connection. Lukács (1971), whose idealization of Bolshevik knowledge production is usually seen as a predecessor of critical theory, did not need Horkheimer's formalistic distinction between "traditional" and "critical" theory because he was still able to refer to the socially concrete distinction between bourgeois academia and Marxist political parties. With the degeneration of the latter (partly due to their own internal contradictions),[2] Horkheimer was left with nothing more than the aspiration for the kind of learning process Lukács had discerned in Bolshevism, divorced from its original but now unacceptable social basis. Given this situation, the ultimate fate of critical theory was academicization (though initially, and significantly, in the form of an independent research institute).

This, however, already introduces a contradiction into the notion (which is also the seed of much unclarity in its exposition). Viewed from this perspective, the idea's center of gravity is precisely the connection to *practice*: rather than being simply *about* reality, knowledge should (also) serve as a direct link in the formation of a *collective will* capable of changing it. For a clear understanding of critical theory, then, it is probably best to begin by reflecting on the changes in the form of knowledge required by this change of intended function. In order to add up to a broader social movement, a diverse series of local social initiatives must not only be unified in theoretical retrospect (or "in their concept," as a Hegelian might say), but also already be developed in reference to a roughly shared interpretation of what is to be done in a shared social environment. The theory must be capable of overcoming the gap between the irreducibly unique situation of each individual actor, and the requirement for collective action. To achieve this—particularly given a globalizing and perpetually reinvented capitalist division of labor—it must enable its readers to interpret their individual situations in relation to their global and historical context, and that requires a representation of that context, of the historical "totality." As such situations are determined by their evolution and systemic environs, such totalization is a precondition of the understanding of any concrete situation. However, such totalization risks imposition of preconstituted conceptual systems on a reality that is necessarily too concrete and complex for any such formal system of concepts.[3]

To avoid this risk, theoretical principles (even the most general, philosophical ones) must be achieved as a posteriori crystallizations of a process of collective experience, which are recognized as in principle open to challenge and revision. Hence, the theory can no longer be identified with the contents of a single text, but rather with the continuous activity of all those involved in a social movement, unified by a division of labor, and whose textual products (of various kinds, of which "metanarratives" are only one) serve only as the way stations of theorization.[4] This combined theoretical effort can therefore express and articulate the interests ordinary people have in their work and lives, and enable their active control over their "lifeworld" by "working up" their tacit technical and social knowledge, thereby facilitating self-management in the workplace and community. It would thereby help to clarify their interconnection in a possible "social plan," subject to less direct forms of democratic control.[5] Such plans would always be revisable, and social science would help to render them evaluable. This conjures the image of a democratic form of socialism, radically distinct from Stalinist practice, but which is also capable of preserving a complex social division of labor and skills.

For this reason, the theory must fulfill a communicative as well as a mimetic function, and orient individuals *to each other*. The theory must therefore be socially positioned as a collective representation, a common reference point for the coordination of diverse actions and a symbol of shared commitment, much as the Bible has been for Christians. This is not a narrowly religious function: social movements require such "bibles" as a condition of organization.[6] However, if the authority of such representations is to be compatible with the rationalistic and democratic ambitions of critical theory, it must again derive from the theory's mode of production/legitimation as a publically examinable and ongoingly revisable vehicle of collective experience, rather than ritual valorization or philosophical demonstration from first principles allegedly so self-evident that their universality does not need to be checked.

However, though academicization may have been an understandable response to the 1920s crisis of European socialism, it is obvious that once academicized this ideal of critical—or, rather, socialist—theory must become a pure utopia, and it is then easy to understand why the commitment to praxis tends to be marginalized, and the commitment to totalization increasingly problematic, for the critical theorists themselves.

To elaborate, a separation of knowledge from nonacademic activity is built into the tacit social contract on which the modern academic system is founded, at least at the level of everyday life, if not always of industry, state, or "national" culture (to which, indeed, it is supposed to contribute in various ways, if indirectly). Academic professions provide a privileged, ordinarily secure, and intellectually rewarding form of what nevertheless, remains at bottom dependent salary work. Having much to lose has encouraged the development of a culture of caution with regards to political activity. Where this culture weakens—usually due to a confluence of a scientific and a broader social interest—the borderline between "scientific" criticism and unacceptable political activism is reasserted (and clarified) through academic freedom cases. Academics genuinely committed to political change generally accept this situation as a Faustian pact, whereby they accept a minimalization of their radical inclinations in their immediate institutional environment in exchange for access to the traditional right/power to research freely, as a means to develop those theoretical analyses that are a necessary condition for effective radicalism throughout the social body as a whole.[7] Indeed, confronting the issue solely as an individual, they have no choice.

Yet, this academic situation places fundamental limits on the kinds of knowledge they can develop. In particular, the ideal of totalization is unattainable for a model of science in which individual production remains a deeply rooted norm. National variations notwithstanding, the concept of tenure basic to the modern research university linked publication to trajectories of career advancement, making individually authored articles or monographs the chief evidential basis for discrimination between individuals for the purposes of appointment and promotion. This has encouraged academics to approach knowledge production as a struggle for distinction in which ideas serve as a proxy for the author's self-advancement, which has in turn encouraged diversification of projects and exaggeration of differences, at the expense of comprehension of the whole.[8] It is enough to reflect that any system of collective representations capable of fulfilling a "totalizing" role must synthesize theorization of, for example, the latest forms of finance capital and the constitution of the working class, with estimates of demographic growth and agricultural and engineering potential, and so forth—to realize that it is quite beyond the efforts of individuals or of small-scale collaboration. At best, such artisanal efforts can produce fragmentary efforts intended to be contributory to such a synthesis—which is quixotic if the synthesis is unrealizable, or if no serious efforts are being made to realize it. Moreover, even if such a theory could be constructed (perhaps with the aid of research assistants) it would only represent the judgment of a small number of academics. Should it prove unsuitable for its intended biblical function, whether through false assumptions, errors of judgment, excessive complexity, and so forth, it would be almost impossible to correct or realign it. Similarly, it would be humanly impossible to keep it up-to-date with new historical developments and critical attacks. In all these ways, it could not lay claim to the rational authority necessary to fulfill its practical functions.

Thus, again, a failure of sociological self-consciousness proves fundamentally debilitating for its capabilities as an objective science. Without some negotiation of these difficulties, and institutional habits, an adequate contextualization of any given social practice is impossible. And without understanding of the values orienting social practices and helping to define their functions within the broader social system, any attempt to reconnect knowledge with "practice" risks loss of critical distance, and cooptation into the agendas of state and capital.[9] This is the main problem with calls for more practical and participatory forms of "action research," which work by getting people talking and thinking together about social interactions they normally conduct unreflectively.[10]

As the process clarifies mutual miscomprehensions and (when successful) builds up a collective normative understanding (or social contract), this can lead directly to voluntary efforts to improve social interaction, and increases in efficiency. However, the attempt to define action research as a formal procedure of *facilitation*, in which the researcher adds no content to local knowledges, disguises the fact that the conclusions of such practices always depend on assumptions about the ends of the particular practice, which may be quite tacit and embodied in the "agenda setting" of the practitioner. The contrast is clearest in industrial action research: where the practitioner is called in by management, the procedure is always oriented to improved efficiency in meeting the productive goals defined by management. The worker is then flattered, by being allowed an element of participation, into collaboration in his own exploitation—a clear case of what Marcuse called "repressive desublimation."[11] Where, on the other hand, the research is conducted by radical socialist unionists, as with the once-famous case of the Lucas aerospace workers,[12] this can involve the development of an alternative production plan developed with reference to the idea of *social need or utility* rather than the needs of capital. This highlights the importance of the shop stewards' broader interpretation of social reality.

It is difficult to imagine academia and unionism being transformed overnight in this direction. Notwithstanding the political barriers to Lucas-type actions, action research is more generally a poor fit with (existing) academic norms: its "reports," such as they are, amount to little more than narrative accounts that are difficult to peer review, and this also makes it difficult to train action researchers. However, if it is often difficult to see how academics might do anything differently, this does not seem the case with the development of a more totalizing approach. Indeed, it seems to me that new technologies may abet such an approach. *If* a group of qualified and socially spirited academics (this would have to mean those sufficiently insured from the demands of the publishing and perishing rat race to which early career academics are subjected) could agree on a series of basic commitments, it would seem possible to develop an e-text or system of e-texts, which could be developed and published gradually, and revised in response to criticism from a broader community of "users," much as open-source software programs are frequently developed in a series of updates.[13]

By throwing their collective authority behind such a project, and by resolving to update, defend it, and perfect it, they would bestow on it a greater cultural significance than any of their individual essays or books.[14] The work would also gain cultural strength from the fact that it could be opened up to continuous criticism—through, for example, a typical Internet discussion forum—by a broader community of interested participants. This would actively involve broader social circles in the active theoretical understanding of social reality, while allowing the theorists access to the insights of the broad panoply of social knowledges and intelligences and, as such, help to maintain the elusive "unity of theory and practice." As such, it could help to refocalize the Left as a political project "beyond the fragments" of various local and single issue struggles.

NOTES

1. Critical theory was one of the original influences on Fuller's (1988) concept of social epistemology. The programmatic distinction between critical and traditional is, of course, Horkheimer's (1999, 188–252). My interpretation of the critical theoretical tradition is influenced by my reading of the work of György Márkus, one of Lukács's former students (see Norrie 2014).

2. We cannot cover this here. However, any interpretation of the matter must proceed in light of Lih's (2008) magnificent critique of both textbook and Trotskyist interpretations of the classical Marxist tradition.

3. This is the main poststructuralist critique of the tradition, the decisiveness of which its historian (Jay 1984) also overemphasises, largely due to Kantian limits on his interpretation of Lukács: in general, the Anglophone commentary industry has done endless mischief in interpreting both Lukács and critical theory, due to its refusal to understand Hegel—though the practical totalization outlined here cannot be identified with Hegel's contemplative, post hoc systematization, and uses different methods. As an additional consideration, if society is considered a distinct level of reality from the social interactions that constitute its microstructure (Bhaskar 1998), then "totalization" at the level of social reality needn't imply knowledge of every social situation.

4. The collective character of theorization, unified by a division of labor, was a standard theme of classical Marxism (Lih 2008).

5. For one attempt to model such a system, see Horvat 1982.

6. The "biblical" metaphor, however, proves all too apt for traditional Marxist parties, which have proved incapable of separating the rational content of their systemic understanding of history from its embedding in Marx's own texts which, accordingly, attain a quasi-sacred status. Indeed, with the (partly unfair) repudiation of "vulgate" writers like Engels and Kautsky, Marxism seems to have passed from a more accommodating "Catholicism" to a more "Protestant" (and sectarian) fundamentalism of the text.

7. Indeed, one function of the name "critical theory" is to accentuate this critical orientation, at the expense of its practical commitments.

8. This does not exclude, but relies on, a parallel process of local synthesis and drawing of connections, which is indeed becoming more fruitful across the sciences. The more diversification there is, the more space there is for such local syntheses, and the more such syntheses are developed, the more each one can serve as the basis for a new disciplinary focus, and fresh diversification of questions.

It might be asked what effect recent transformations of the research university (see, e.g., Ziman 1994) have on this picture. On the one hand, they seem to make little difference. The life of the academic remains a comparatively satisfying one, and traditional academic values and self-perceptions retain an (increasingly embattled) appeal (Bryson 2004). Moreover, state-imposed accounting systems intended to increase academic productivity only increase the centrifugal trajectory of academic publishing. At a deeper level, though, an increasing tendency to interdisciplinarity, the increasing cultural reach of academic knowledges, and even the connection with industry itself, might be seen as pressures against the individualist norm.

9. This is even true where the client is a communal, rather than a public or private institution, as was unintentionally revealed by David Cameron's by now notorious call for a "Big Society" in which the state's welfare provision is loaded onto the charity and voluntarism of private individuals, allowing the channelling of taxation to service capital accumulation.

10. Carr and Kemmis (1986) associated such forms of knowledge with Habermas's concept of critical theory, though somehow without mentioning *capitalism* once. Arguably, Dewey's pragmatism is a more natural (and more common) philosophical fit for such abstracted models of learning.

11. Marcuse 1991.

12. Wainwright and Elliott 1982.

13. The series of basic commitments would have to include agreement on a method of systematization, as well as on a number of important substantive issues (e.g., positions on alternative technology, on the disaster of Stalinist "socialism"). To be sure, some might consider coming to such an agreement the main problem! However, all that is required is that a small group of collaborators come to a broad agreement—those who disagree can form their own competing collective!

14. The computer already allows a distinct way of writing from that which produced most of the canonical texts of contemporary discussion, which were either typed or hand written and then typed up by someone else. In the earlier mode, writing must follow the flow of thought, and thus is less distinguishable from speech, and the articulation of a subjective viewpoint. Computer writing, on the other hand, has

divorced text production from speech, allowing a greater degree of objectification (and perhaps potentially alienation) *vis-à-vis* its author. By analogy to the "slow cooking" movement, we might say the computer allows "slow writing," and with it, a greater potential for attention to systemic form. Indeed, if academia is judged by the production of knowledge rather than texts, it seems to me a straightforward case of relations of production retarding forces of production that such "perfectionism" is today seen as a vice which (ironically) "retards" the academic production of junk knowledge.

BIBLIOGRAPHY

Bhaskar, Roy. *The Possibility of Naturalism: A Philosophical Critique of the Contemporary Human Sciences*. London: Routledge, 1998.
Bourdieu, Pierre. *Science of Science and Reflexivity*. Cambridge: Polity Press, 2004.
Carr, Wilfred and Stephen Kemmis. *Becoming Critical: Education, Knowledge and Action Research*. London: The Falmer Press, 1986.
Durkheim, Emile. *Rules of Sociological Method*. New York: Free Press, 2014.
Foucault, Michel. *The Order of Things. Archeology of the Human Sciences*. New York: Routledge, 2001.
Fuller, Steve. *Social Epistemology*. Bloomington: Indiana University Press, 1988.
Horkheimer, Max. *Critical Theory*. New York: Continuum, 1999.
Horvat, Branko. *The Political Economy of Socialism: A Marxist Social Theory*. Oxford: Robertson, 1982.
Jay, Martin. *Marxism and Totality*. Cambridge: Polity Press, 1984.
Lecourt, Dominique. *Pour une critique de l'épistémologie (Bachelard, Canguilhem, Foucault)*. Paris: François Maspero, 1972.
Lih, Lars. *Lenin Rediscovered*. Chicago: Haymarket, 2008.
Lukács, Georg. *History and Class Consciousness*. London: The Merlin Press, 1971.
Marcuse, Herbert. *One-Dimensional Man*. London: Routledge, 1991.
Norrie, Stephen. "Towards a Critical Theory of High Culture: The Work of György Márkus." *Journal of Critical Realism* 13, no. 5 (2014): 467–97.
Wainwright, Hilary and David Elliott. *The Lucas Plan: A New Trade Unionism in the Making?* London: Allison & Busby, 1982.
Ziman, John. *Prometheus Bound*. Cambridge: Cambridge University Press, 1994.

Chapter 6

The Politics of Social Epistemology

Susan Dieleman, María G. Navarro, and Elisabeth Simbürger

For the last twenty-five years, those with an interest in social epistemology sooner or later realized, perhaps with some amazement, that the label "social epistemology" actually includes two varieties. For the untrained eye, it may initially take some effort to figure out the differences between the research program, "social epistemology," as set out by Steve Fuller twenty-five years ago in the journal *Social Epistemology* (founded 1987) and a monograph of the same title (1988), and Alvin Goldman's research program, "social epistemology," developed shortly thereafter, with the publication of *Knowledge in a Social World* (1999) and the journal *Episteme: A Journal of Individual and Social Epistemology* (founded 2004). However, as has been demonstrated elsewhere, these seemingly identical research programs are in fact worlds apart (Collin 2013; Remedios 2013; Vähämaa 2013). At base, the disputes between Fuller's *sociological* social epistemology and Goldman's *analytic* social epistemology seem to depend upon a diametrically opposed understanding of the role of normativity in social epistemology.

Fuller's understanding of social epistemology stands in contrast to analytic social epistemology, which he characterizes as a

> more 'classical' approach of starting from the individual knower who is ultimately concerned with whether her beliefs correspond to an epistemic standard that is presumed to exist independently of her own individual or collective activity—be that standard cast in supernatural (e.g., the Cartesian deity) or naturalistic (e.g., the Quinean physical environment) terms. (Fuller 2012, 276)

For Goldman, the debate is characterized somewhat differently. He writes,

> According to one perspective, social epistemology is a branch of traditional epistemology that studies epistemic properties of individuals that arise from

> their relations to others, as well as epistemic properties of groups or social systems. . . . A very different perspective would associate 'social epistemology' with movements in postmodernism, social studies of science, or cultural studies that aim to replace traditional epistemology with radically different questions, premises, or procedures. (Goldman 2010, 1)

Goldman goes on to suggest that analytic social epistemology is "real" epistemology, while Fuller's version of social epistemology—which we will subsequently refer to as *sociological* social epistemology—is "not part of epistemology at all" (Goldman 2010, 1).

We have two main objectives in this chapter. The first is to suggest that social epistemologists tend to overlook or underplay the importance of the political dimensions of knowledge production. The second is to demonstrate the importance of politics to social epistemology using three cases: knowledge production in social movements, in political campaigns, and in the university. We will conclude by highlighting how the study and practice of social epistemology might look different were its political dimension to be more centrally located.

The political dimension of knowledge production is twofold. As social epistemologists motivated by the politics of knowledge production, we start from the assumption that there is no such thing as neutral knowledge or neutral knowledge producers. Knowledge always goes hand in hand with normative standpoints. On the other hand, the political dimension of knowledge production cannot just be confined to a normative position with regard to the knowledge we produce. It also involves taking normative positions in and around the contexts of knowledge production, such as the university, public policy, social movements, and think tanks, to name a few possible sites.

Of course, to suggest that social epistemology tends to ignore politics is not to say that neither the sociological nor the analytic version of social epistemology is altogether oblivious to the political dimensions of our knowledge practices. Analytic social epistemology, for example, has benefited from careful analyses of the role that power plays in the attribution of expertise and when one is justified in trusting another's testimony, and efforts to develop and evaluate accounts of collective epistemic agency and responsibility are clearly political in nature, or at least in effect. Yet its tendency to neglect the fundamentally social nature of the *contexts* in which expertise, epistemic agency, and responsibility, and so on, are practiced undermines its own breadth of applicability. Vähämaa, for example, argues that, whereas sociological social epistemology dedicates itself to the study of socially shared beliefs and how they are understood by communities, for analytic social epistemologists, "'real' knowledge is constrained by propositional logic, which is derived from language and is constructed in social settings" (Vähämaa

2013, 36). Vähämaa criticizes this view for its tendency to leave aside the important role that social groups and social practices play in the construction and transformation of knowledge. As a result, one might be left wondering whether the notion of the social takes more than a merely decorative place in the description of the research program, as it clearly comes in second place after epistemology.

Of course, sociological social epistemology is more attentive to the normative dimension of knowledge production. As Fuller set out in his program in 1988, one of the key questions is how the pursuit of knowledge *ought* to be organized. The underlying thesis is that the production of knowledge is a normative endeavor. This perspective has been disputed ever since by science and technology studies of the Latourian kind that opts for a non-normative approach to knowledge production, as is well documented in Steve Fuller's and Bruno Latour's respective controversies (Barron 2003). However, despite being more normative than its analytic counterpart, and more attentive to the institutional structures that have determined the limits of knowledge historically, sociological social epistemology tends to pay insufficient attention to the role that power, domination, subordination, and oppression play in the historical creation of norms and practices and systems of knowledge. For example, sociological social epistemology has largely seemed to ignore the so-called postcolonial revolution in the social sciences and humanities (Bhambra 2007; Connell 2007) and has paid little attention to current discourses on the social sciences and humanities in a global context and the possibilities and challenges of crafting connected histories and sociologies that go beyond the separation between the so-called global North and South (Bhambra 2014; Keim et al. 2014). Moreover, one of the key deficiencies of sociological social epistemology and the work of its adherents is that empirical work remains to be the exception rather than the rule. This is not to overstate the importance of empirical work or to undermine the significance of theoretical work. Nevertheless, it is the neglect of the empirical dimension of social reality that finally bites social epistemology the hardest and occasionally reduces social epistemology's selling point of being political to a mere banner. To put it differently, whereas sociological social epistemology emphasizes the importance of normativity in its research program and distinguishes itself from analytic social epistemology in this respect, there may still be some mileage to living up to this promise in a more encompassing way.

In what follows, we provide an account of three distinctly political issues that we think social epistemologists should pay attention to: the production of knowledge in social movements, in political campaigns, and in the university. Each of these sites is *political*, we suggest, insofar as it requires that social epistemologist engage with and take up a normative position regarding the site of knowledge production.

SOCIAL EPISTEMOLOGY AS PUBLIC PHILOSOPHY: KNOWLEDGE PRODUCTION IN SOCIAL MOVEMENTS, SUSAN DIELEMAN

A political social epistemology is necessarily attentive to the institutional structures that make knowledge possible—or, more accurately, make some knowledges possible while rendering other knowledges impossible. Indeed, this seems to be the fundamental feature of institutions: they function as sites of legitimation. Disciplinary boundaries are determined according to what counts as the right sort of knowledge to count as a contribution. Disciplinary boundaries tend to be porous in reality, but they remain heavily policed nonetheless—one need only look at top-tier journals in a particular discipline (and I have my own discipline—philosophy—in mind) to see where these boundaries are drawn, and how institutions legitimate their own knowledges. However—and this is, of course, not an altogether novel claim—the danger that follows from institutionalization is that it tends toward the maintenance of the status quo, and reduces the chances of epistemic novelty.

To change knowledge—to shift the boundaries of what counts as knowledge and what does not, and to reconsider who counts as a knower and who does not—requires that we attend to the institutional settings in which knowledge is produced and legitimated. It is particularly incumbent on those who inhabit or populate institutions, if they are unsatisfied with the status quo (and I want to suggest that they have good reason to be unsatisfied), to challenge those very institutions. This might be a somewhat paradoxical demand—to ask that those whose work constitutes and is legitimated by an institution attend to the costs of that legitimation. But there are benefits to doing so. Indeed, if only certain knowledges and their related practices are considered legitimate within a disciplinary boundary, then one is justified in asking "What knowledges and practices are being left out? Is it the case that, simply by virtue being left out, they are neither knowledges nor practices of knowledge?" I suggest that the social epistemologist has a responsibility—an *epistemic* responsibility—to seek out alternative knowledges and practices of knowledge that are, because of the political practice of disciplinary boundary setting, not traditionally recognized.

In her 2012 paper "How Is This Paper Philosophy?" Kristie Dotson makes a similar point. She writes,

> The environment of professional philosophy manifests symptoms of a culture of justification, i.e. a culture that privileges legitimation according to presumed commonly-held, univocally relevant justifying norms, which serves to amplify already existing practices of exceptionalism and senses of incongruence within the profession. (Dotson 2012, 6)

In other words, the discipline of philosophy legitimates its own knowledges and knowledge practices, thereby maintaining them. Dotson goes on to recommend that a culture of praxis take the place of a culture of justification within the discipline of philosophy. A culture of praxis, she argues, would have the following features:

> (1) value placed on seeking issues and circumstances pertinent to our living, where one maintains a healthy appreciation for the differing issues that will emerge as pertinent among different populations and (2) recognition and encouragement of multiple canons and multiple ways of understanding disciplinary validation. (Dotson 2012, 17)

I want to suggest that the culture of praxis Dotson recommends is similar to my own recommendation that social epistemology ought to be public philosophy.

A useful model here is the idea of theorist or philosopher as "rearguard," a view that can be found in Linda Martín Alcoff's reading of Enrique Dussel, whose philosophy of liberation, she suggests, "invokes the idea of philosophers as analytical transcribers or rear-guard theorists, not inventors or originators so much as those who give philosophical articulation to the ideas embedded in the praxis and lived experience of the activist oppressed" (Alcoff 2012, 62). Public philosophy, then, is best understood as the engagement in and contribution to activist groups on the part of philosophers, whose conceptual tools and analytical skills might serve the liberatory ends set out by others. This suggestion—that the social epistemologist is a public philosopher—amounts to asking the following question: Which social is the subject and the site of social epistemology? Social epistemologists should be among those leading the challenge to broaden and alter the social we, as social epistemologists, take as the topic for and site of our engagement. Social epistemology should be politicized; it should be public philosophy.

SOCIAL EPISTEMOLOGY: FACING THE CHALLENGE OF KNOWLEDGE PRODUCTION IN POLITICAL CAMPAIGNS, MARÍA G. NAVARRO

From the dynamics of electoral campaigns, it can be inferred that some of the most important producers of social meanings are the political parties and the leaders or candidates representing them. Each electoral campaign is different and each one conditions in a different way the type of strategy needed to establish the encounter between citizens' knowledge (i.e., known information) and the information that is discussed in the candidates' messages (i.e., new information).

The elementary classification of campaign types by Newman and Sheth (1987) can be useful to introduce the debate about the theoretical space of opportunity of social epistemology in the analysis of knowledge production in electoral campaigns. It is well known that Newman and Sheth distinguish up to four types of campaigns. These are the result of the combination of two factors: the nature of the implication in the election (which can be high or low) and the degree of familiarity that the elector establishes with the candidates (the intensity of which varies from high to low levels). In my view, the very different kinds of campaigns that can take place, as well as the idea that there is a space of encounter between known and new information, defy the thesis according to which the beliefs of the electors correspond to standards that exist independently of individual or collective action. The electors experience not just the very different kinds of pressures that operate during a campaign (e.g., institutional pressure, pressures derived from the human natural disposition to influence and being influenced), they also experience diverse degrees of implication and they can believe that certain elections may be decisive for their future, even when they have a scarce knowledge of the leaders.

The question I would like to address is simple: if we take into account that for the analytic social epistemology defended by Goldman (2011, 11–37) it is not enough to examine the social contexts of believing, as it is necessary to illuminate the conditions of epistemic success or failure, then how could this dichotomic model of analysis be applied to the effects produced by the electoral campaigns as they are classified by Lazarsfeld (1968) (e.g., the campaigns would activate the political predispositions of the electors, they would reinforce the vote intentions, and in some cases they could produce processes of conversion or change in vote intention) without eliminating de facto the phenomenon that one intends to analyze?

My impression is that when we suppose that there might be a level of epistemic analysis independent from the individual or collective activity that is being examined, we only obtain the annulation of social reality as a result because we take it to a simple diagnosis about its supposed epistemic failure and/or success. It seems reasonable to think that a political conception of social epistemology is needed to analyze the political attitudes and predispositions of the citizens during the periods of electoral campaigns. It can be said that in those moments of institutional tension there are variables that determine the vote orientation that do not depend only on known information but also on information that one acquires in the process. And that the phenomenon of collective epistemic agency cannot be evaluated without assuming that it develops and configures itself in a political context. The type of communication that takes place during electoral campaigns defies the analytic conception of social epistemology. It would be even necessary to eliminate the concepts coined by political science to finally capture effects produced during the development of an electoral campaign (Brady et al. 2009) that do not contradict the normative dimension as it is assumed by analytic social epistemology.

But the limitations of social epistemology in general (i.e., in its sociological and analytical projections) when applied to the field of politics and, particularly, when applied to the complex processes of public opinion conformation and to the production of knowledge during election campaigns, are similar in the case of big data, a fundamental promise in the future economic revolution as concluded from the reports of International Data Corporation (Schmarzo 2013; Ohlhorst 2012), and a challenge in epistemic terms. The world produces more and more information, via more devices, in more sites and using more apps. According to the European Union, every minute enough information is generated to be recorded in 360,000 DVDs. How would those unstructured data affect the processes of belief acquisition, and that of the public opinion conformation, when "putting some brain to them" is achieved? A challenge for social epistemology is to explain how and who will transform big data into knowledge during political campaigns, and what will be, in each case, the political consequences of the new form of epistemic agency that dawns on the horizon.

SOCIAL EPISTEMOLOGY'S GAZE AT ITSELF: KNOWLEDGE PRODUCTION IN THE UNIVERSITY, ELISABETH SIMBÜRGER

Even within sociological social epistemology itself, the calling for normativity with regard to knowledge production does not always seem to translate into more tangible, political realities. It is a worthwhile endeavor shedding a bit more light on the question of social epistemology's dedication to politics on the ground, focusing on the university as a workplace and as a site of neoliberalism. Are we really political? Do we engage with the materiality of what knowledge production is about?

For this purpose of gazing at ourselves, I will turn to Alvin Gouldner and his work on reflexivity in the social sciences. In *The Coming Crisis of Western Sociology*, Alvin Gouldner (1970) analyzes the intertwining of social theory and practice with the political surroundings at the time in the 1960s and 1970s. Similarly to C. Wright Mills, he suggests that being a sociologist is a life-encompassing activity that cannot be discarded at the doors of a university. As such, Gouldner already presents his social epistemology business card. The process of awareness of ourselves in our totality in relation to our research and the outside world is at the core of Gouldner's Reflexive Sociology program. What makes his work so distinctive is that it is an epistemological position with practical and political implications. For him, critique can never be a static undertaking and needs to be continually revisited (Gouldner 1970) as he has shown in his work about the growing convergence between Functionalism and Marxism. Being a Marxist himself, he accused Marxist sociologists of being in a static relationship with their theory, of not living up to their strong theoretical claims of critique and not questioning the foundations of their thought (Gouldner 1970). In fact, Gouldner saw open-mindedness for hostile information as a strategy that prevents us from getting lost in dogmatic thought. As should be clear by now, Gouldner can be seen as a true social epistemologist.

Without doubt representatives of Fullerian social epistemology have repeatedly shown that they are open-minded to seemingly hostile information, often playing devil's advocate (Fuller 2007; Fuller and Lipinska 2014). However, there seems to be less documentation for social epistemology's commitment in conceptual terms—speaking of Fullerian social epistemology as a collective—to politics within the university on a global scale and against the prevalent conditions of knowledge production within higher education. Academic capitalism is everywhere (Münch 2011). At the heart of the transformation of the university is a transformation of its labor relations and, as a consequence, a transformation of the kinds of knowledge that we produce (Roggero 2011).

My general concern is not that social epistemology's normativity with regard to these quests was totally absent. Yet, there is a tendency in the social sciences and humanities—and this also applies to this collective—to not name things or to disguise their materiality by choosing alternative labels. However, labeling does have a performative effect after all. The labeling of things becomes part of the materiality of an object after a while. "Social practices" and "producing the social" have become the new buzzwords of our time (Ariztía 2012; Camic et al. 2011). As argued elsewhere, as important the latest "turn to practice" is to study the production of the social sciences, it seems to systematically leave aside one dimension that is crucial in the shaping of practices both in the social sciences and in the sciences: the labor relations of academic knowledge production (Simbürger 2014).

Of course there are exceptions. To take an example Stephen Norrie, from the Social Epistemology Review and Reply Collective, enriched the debate on academic work by providing us with creative approaches on how to form an academic collective in spite of institutional pressure to publish individually in A-rated journals, constraints that are especially prevalent for early career academics in precarious working conditions (2011).

This is not to say that the academic work carried out within the collective is not an act of resistance of some sort. Being cynical, one could argue that taking the time to write a short piece for the online collective and thus having less time left to write for an A-rated journal, could in times of pressure for output already be considered as a postmodern act of solidarity, albeit in quite disguised ways. However, naming and labeling—talking about academic work—is the first step. Perhaps every now and then it is necessary to think and act beyond the neoliberal university.

CONCLUSION

To the extent that *analytical* social epistemology remains mired in debates that have characterized more traditional approaches to epistemology, its political dimension remains underdeveloped. On the other hand, whereas *sociological* social epistemology is very explicit about its normative position, its focus is limited to formal sites of knowledge production rather than informal sites, thereby ignoring sites of social epistemology other than universities and think tanks. In this chapter, we have recommended that this political dimension be made more central to the study and practice of social epistemology. Moreover, if social epistemology is to live up to its political commitments, it is important to prevent its becoming an inward-looking group, remaining open toward schools of thought and ideas that may not necessarily be called social epistemology but may yet reflect like-minded ideas and approaches.

BIBLIOGRAPHY

Alcoff, Linda Martín. "Enrique Dussel's Transmodernism." *Transmodernity: Journal of Peripheral Cultural Production of the Luso-Hispanic World* 1, no. 3 (2012): 60–68.

Ariztía, Tomás, ed. *Produciendo Lo Social. Usos de las Ciencias Sociales en el Chile Reciente*. Santiago de Chile: Ediciones Universidad Diego Portales, 2012.

Barron, Colin. "A Strong Distinction Between Humans and Non-Humans is no Longer Required for Research Purposes: A Debate Between Bruno Latour and Steve Fuller." *History of the Human Sciences* 16, no. 2 (2003): 77–99.

Bhambra, Gurminder K. *Rethinking Modernity: Postcolonialism and the Sociological Imagination.* Basingstoke: Palgrave, 2007.

Bhambra, Gurminder K. *Connected Sociologies.* London: Bloomsbury, 2014.

Brady, Henry E., Richard Johnston, and John Sides. "The Study of Political Campaigns." In *Capturing Campaign Effects,* edited by Henry E. Brady and Richard Johnston, 1–26. Michigan: University of Michigan Press, 2009.

Camic, Charles, Neil Gross, and Michele Lamont, eds. *Social Knowledge in the Making*. Chicago: Chicago University Press, 2011.

Collin, Finn. "Two Kinds of Social Epistemology." *Social Epistemology Review and Reply Collective* 2, no. 8 (2013): 79–104.

Connell, Raewyn. *Southern Theory: The Global Dynamics of Knowledge in Social Science.* Cambridge: Polity Press, 2007.

Dieleman, Susan. "Social Epistemology as Public Philosophy." *Social Epistemology Review and Reply Collective* 2, no. 12 (2013): 69–71.

Dotson, Kristie. "How is This Paper Philosophy?" *Comparative Philosophy* 3, no. 1 (2012): 3–29.

Fuller, Steve. *Social Epistemology.* Indiana: Indiana University Press, 1988.

Fuller, Steve. *Science vs. Religion? Intelligent Design and the Problem of Evolution.* London: Polity, 2007.

Fuller, Steve. "Social Epistemology: A Quarter-Century Itinerary." *Social Epistemology* 26, nos 3–4 (2012): 267–83.

Fuller, Steve and Veronika Lipinska. *The Proactionary Imperative: A Foundation for Transhumanism.* London: Palgrave Macmillan, 2014.

Goldman, Alvin. "Why Social Epistemology is *Real* Epistemology." In *Social Epistemology*, edited by Adrian Haddock, Alan Millar, and Duncan Pritchard, 1–28. New York and Oxford: Oxford University Press, 2010.

Goldman, Alvin. "A Guide to Social Epistemology." In *Social Epistemology: Essential Readings*, edited by Alvin I. Goldman and Dennis Whitcomb, 11–37. Oxford: Oxford University Press, 2011.

Gouldner, Alvin Ward. *The Coming Crisis of Western Sociology.* London: Heinemann, 1970.

Keim, Wiebke, Ercüment Çelik, Christian Ersche, and Veronika Wöhrer, eds. *Global Knowledge Production in the Social Sciences. Made in Circulation.* London: Ashgate, 2014.

Lazarsfeld, Paul F., Bernard Berelson, and Hazel Gauder. *The People's Choice.* New York: Columbia University Press, 1968.

Münch, Richard. *Akademischer Kapitalismus. Über die politische Ökonomie der Hochschulreform.* Berlin: Suhrkamp Verlag, 2011.

Newman, Bruce I. and Jagdish N. Sheth. *A Theory of Political Choice Behavior.* New York: Praeger Publisher, 1987.

Norrie, Stephen. "Three Social Contracts for an Academic Collective." *Social Epistemology Review and Reply Collective* 1, no. 1 (2011): 14–24.

Ohlhorst, Frank. *Big Data Analytics: Turning Big Data into Big Money.* New York: John Wiley & Sons, 2012.

Remedios, Francis. "Orienting Social Epistemology." *Social Epistemology Review and Reply Collective* 2, no. 8 (2013): 54–59.

Roggero, Gigi. *The Production of Living Knowledge. The Crisis of the University and the Transformation of Labor in Europe and North America.* Philadelphia: Temple University Press, 2011.

Schmarzo, Bill. *Big Data: Understanding How Data Powers Big Business.* New York: John Wiley & Sons, 2013.

Simbürger, Elisabeth. "The Labor of Knowledge in the Making of the Social Sciences." *International Sociology Reviews* 29, no. 2 (2014a): 89–97.

Simbürger, Elisabeth. "Social Epistemology as Work: A Quest for Normativity at the University." *Social Epistemology Review and Reply Collective* 3, no. 4 (2014b): 4–6.

Vähämaa, Miika. "Secrets, Errors and Mathematics: Reconsidering the Role of Groups in Social Epistemology." *Social Epistemology Review and Reply Collective* 2, no. 9 (2013): 36–51.

Part II

EXTENDING CONCEPTIONS OF KNOWING

Chapter 7

Metaphor and Social Epistemology

Martin Evenden

Metaphor is important to social epistemology in a number of ways. It is pervasive and crucial in the direction, development, and understanding of scientific theories (e.g., cognitive psychology largely understands the brain/mind as a computer), has been used by scientists in conjunction with the media to exacerbate a climate of fear (e.g., bird flu as a natural bioterrorist) in order to increase resource allocations from policy makers (Nerlich and Halliday 2007) and as tools of persuasion (e.g., the genome as a map) when scientists want to promote the value and social meaning of their science to the public (Nelkin 2001). Accordingly, "Metaphors have profound influences on how we conceptualize and act with respect to important societal issues" (Thibodeau and Boroditsky 2011). Metaphoric self-consciousness can counteract dogmatic tendencies in science, and help the public distinguish political claims from their scientistic clothing.

WHAT IS METAPHOR?

Despite metaphor's traditional associations with anomaly and ornamental use in aesthetics and literary language, Lakoff and Johnson (1980a, 1999) have convincingly argued it is instrumental to everyday language and thought in ways people do not often realize. Their central claim is that metaphor operates according to the formula A is B at the cognitive level and systematically motivates corresponding linguistic expressions. For instance, when we use expressions such as being "lost," "following a path," or "going places" to discuss important life experiences, they are held to be linguistic manifestations of the underlying conceptual metaphor, "A purposeful life is a journey," which is understood as a mapping between two conceptual domains:

the source domain (a journey) and a target domain (a purposeful life). What transpires here is that we use something that we have direct physical or social experience of, a journey (the source domain), to provide structure to something that is more abstract and does not have clearly defined boundaries, a purposeful life (the target domain), thereby making it easier to understand. As such, metaphor is among the most basic mechanisms we have for understanding our experience.

Nonetheless, while conceptual metaphor theory revolutionized and is currently the dominant paradigm in metaphor studies, its focus is on the cognitive aspects of metaphor from which language is held to emerge. When analyzing how metaphors are used in discourse, it does not readily account for context and speaker intention; identifying the conceptual metaphors underlying a text does not in itself show us *why* they have been chosen or take into consideration their relation to the textual genres they appear in. Consequently, in addition to its linguistic and cognitive dimensions, Charteris-Black (2004) argues that "a complete theory of metaphor must also incorporate a pragmatic perspective that interprets metaphor choice with reference to the *purposes of use* within *specific discourse contexts*." Accordingly, this chapter will discuss metaphor's situated applications in terms of both the cognitive and pragmatic roles they perform in relation to social epistemology.

METAPHOR AND THEORY CONSTRUCTION

Metaphor is an essential source of insight for scientists in their efforts to understand the workings of nature, which are often complex and inaccessible to sensory experience (Semino 2008, 131–32). Their visualizable character plays an important role in creating new theories and changing perspectives through the provision of conceptual and lexical resources.

In recognition of this, Boyd (1993) has indicated the need to distinguish between two categories of metaphor that fulfill different epistemological roles in science and theory change. The first:

> which might be plausibly termed exegetical or pedagogical metaphors, play a role in the teaching or explication of theories which already admit of entirely adequate nonmetaphorical (or at any rate, less metaphorical) formulations. I have in mind, for example, talk about "worm-holes" in general relativity, the description of the spatial localization of bound electrons in terms of an "electron cloud," or the description of atoms as "miniature solar systems." (485–86)

While Boyd permits such metaphors the possibility of playing an important role in theory change, for example, helping to recruit scientists in establishing new paradigms, he maintains their role in theory construction is limited because they can be replaced by paraphrases describing the same phenomena without loss of information: thus, they are held to merely describe existing knowledge.

For Boyd (1993), more interesting and more important from a scientific perspective are theory-constitutive metaphors:

> The cases of scientific metaphor . . . [whose] metaphorical expressions constitute, at least for a time, an irreplaceable part of the linguistic machinery of a scientific theory . . . which scientists use in expressing theoretical claims for which no literal paraphrase is known. Such metaphors are *constitutive* of the theories they express rather than merely exegetical. (486, original emphasis)

It is this inability of such metaphors to be paraphrased that makes them more than mere heuristics and as such, unique components that are integral in the development and articulation of theory construction. This is because of the "catachretic" role they perform: catachresis here refers to the introduction of scientific vocabulary where none previously existed, "for features of the world whose existence seems probable, but many of whose fundamental properties have yet to be discovered" (489–90).

To illustrate these claims, Boyd draws on the "brain is a computer" metaphor predominant in cognitive psychology and the matrix of spin-offs it has produced that has assisted in organizing its investigation. For instance, "information processing," "computations," and "memory storage capacities" are expressions used to explain the operations of the brain that presently have an indispensable role in cognitive psychological theorizing. The very intelligibility of these terms is dependent on their relation to the computer metaphor and there are no paraphrases psychologists can offer to draw the same distinctions.

Nonetheless, this category distinction between theory-constitutive and pedagogical metaphors is slipperier than Boyd suggests, as various commentators have shown (Bucchi 1998; Knudsen 2003; Semino 2008). Metaphors that are used for illustrative purposes in nonspecialist contexts and genres are often also the same ones used to develop conceptual foundations in specialist ones. Another problem is that the status and meaning of metaphorical concepts changes as a theory develops over time. While theory-constitutive metaphors are introduced to understand poorly understood subject matter, after sustained periods of clarification and testing, they become increasingly invisible and "closed," whereby the scientists using them no longer consider them metaphorical and use them like any other scientific concepts (Knudsen 2003, 1254).

Thus, while these category distinctions may be analytically useful, it is more accurate to say they describe functions of metaphors rather than exclusive properties belonging to them. This does not mean it is not possible for a particular metaphor to be used for theory-constitutive or pedagogical purposes only, as the paraphraseability criterion suggests, but "identical metaphorical expressions are capable of serving several purposes, dependent on context" (Knudsen 2003, 1259).

METAPHOR AS HIGHLIGHTING AND HIDING

It is important to be aware that in addition to the insights they provide, metaphors also constrain given that in the process of foregrounding certain elements of the target domain, their internal structure necessarily conceals others. Yob (2003) has noted this is a problem inherent in all metaphors:

> Since in a sense metaphors are an artifice, a tool, for opening up possible conceptual territories for explanation, their connections and dynamics in constructing knowledge have inherent limitations. Primarily, a metaphor is not the thing being referred to but a symbol of it. If it were the same as the thing it was referring to it would not be needed. Therefore, it is other than and in some respects less than what it refers to, even when referring powerfully and provocatively. (133–34)

Hence, while thinking about the brain and by extension the mind in terms of computers has been productive and fruitful, particularly in areas focusing on understanding and retaining information, it has limitations that some involved in cognitive science would do well to be aware of. For instance, the computer frame does not easily fit into it other elements that clearly affect cognition, such as emotions, bodily experience, social relations, and context (Semino 2008, 136). By understanding the mind as a container, it views thinking as something that occurs exclusively within the boundaries of the head. The problem is not using metaphor to try and understand something; it is simply that the mind is not a computer; thinking about it in this way alone will therefore constrain our understanding of it and inhibit new ideas and developments that might emerge if we also conceptualized the mind in other ways.

Different metaphors for a given phenomenon invite different sets of questions that scientific theories attempt to address, and as Sternberg (1990) observes, "We often see theories as competing, when in fact they are not: They are different answers to different questions, not different answers to the same question" (284). The theory of the nature of light can offer useful guidance as to how different metaphors might successfully be combined to address different questions in contemporary cognitive science and other scientific domains: "sometimes it behaves like a particle; and sometimes it behaves like a wave. Neither analogy by itself captures all the phenomena displayed by light, but both are extremely useful in characterizing some of those phenomena" (Eliasmith 2003, 495).

If no single metaphor is going to give us all the answers due to the aspects of reality it hides, it makes sense to use different metaphors that offer different insights toward understanding the same concept. Lakoff and Johnson (1980b) claim that scientists generally hold an aversion toward using alternative metaphors because of a well-intentioned insistence on consistency and suggest that while consistency may indeed be desirable, there are times when scientific understanding would be better served by making use of alternative metaphors.

An interesting alternative that allows new understandings of mind to emerge by moving it out of the head and blurring common distinctions such as environment/individual, inside/outside, and self/other is a rhizome (Duffy and Cunningham 1996, 177). A rhizome, such as a tulip or ginger, is a root crop that consists of a mass of stems, roots, and fibers whose fruits are tubers, bulbs, and leaves. By using a labyrinthine root structure to explore how mind can be understood, this metaphor suggests characteristics that are

> dynamic, continually growing in dimension in which there is potential for infinite connection among heterogeneous features that are not necessarily hierarchically organized, but are in a continuous interconnected system that cannot be ruptured, something that has no inside or outside, yet has multiple entrances. (Schuh and Cunningham 2004, 338)

While a rhizome itself as a root crop is easy to visualize, the concept of mind it represents is very abstract. The key point to understand is that a rhizome exists in multiple locations; thus, its fundamental principle is based on connectivity. Its nexus-like structure represents mind and knowing as distributed at the systems level, encompassing processes that contribute to cognition wherever they occur, reflecting its multidimensional nature. Correspondingly, the mind is distributed in social, cultural, historical, and institutional contexts: "Thinking, or whatever we choose to call the activity of mind, is always dialogic, connected to another, either directly as in some communicative action or indirectly via some form of semiotic mediation: signs and/or tools appropriated from the sociocultural context" (Duffy and Cunningham 1996, 177).

Understanding the mind as a network distributed in many contexts provides an interpretation of mind more effective than a container-based metaphor when trying to conceive of embodied thought processes embedded within sociocultural (e.g., educational) contexts with supporting artifacts. If cognition is understood as a systemic arrangement, it enhances possibilities of extending our cognitive potential by refining our ability to creatively rearrange our environment in order to scaffold our thinking and activities (Pea 1993).

Nonetheless, theorizing internal states remains important and there are still times when it is more profitable to understand the mind as a discrete entity located inside the head, which focuses on a given set of salient features: a holistic metaphor that highlights too much of the target domain can, in doing so, also hide important features.

Various metaphors for the mind, such as the blank slate and telephone switchboard, have been used over time and discarded in favor of more successful ones, and in future, new metaphors will emerge and others may even blend (Sternberg 1990, 285).

METAPHOR AND PERSUASION

This highlighting and hiding aspect of metaphor also reveals how using metaphors to understand something is not always neutral or benign. Its capacity to shape how we perceive a target domain via a lens like effect not only directs our focus toward specific features, but often results in a metaphoric image that carries implicit moral and social evaluations, which can be exploited for ideological purposes. Thus, metaphor has a "framing" or persuasive function in addition to its constitutive and pedagogical functions that has the potential to covertly influence our reasoning and arouse our emotions.

While the persuasive power of metaphor has long been recognized, at least among politicians (Ball 2011), it has recently been empirically demonstrated by Thibodeau and Boroditsky (2011). They demonstrated that people systematically offered different solutions to crime depending on whether it was framed as a virus or beast ravaging a city. Those who read a report about increasing crime rates with crime framed as a virus proposed solutions such as investigating the root causes of the issue and enacting social reforms; those who read the same report with crime framed as a beast focused on methods of law enforcement and punishment. Furthermore, it was found people did not recognize the influence of the metaphors in their decision making; instead, they identified the statistics in the crime report as motivating their decisions.

Ball (2011) notes,

> It isn't hard to see why 'crime as wild beast of prey' encourages people to think about how to cage or kill it, whereas 'crime as virus' fosters more eagerness for 'scientific' understanding of causes. . . . In both these cases, crime is presented as a (malevolent) force of nature, outside human agency. Whether beast or virus, the criminal is not like us—is not human. (www.nature.com/news/2011/230211/full/news.2011.115.html)

The ability to influence how we perceive social realities clearly makes metaphor a powerful communicative tool in public debates on the importance of scientific research. Scientists through the media have used metaphors to generate both positive expectations, such as when geneticists have referred to the genome as a map to promote the diagnostic value of their work as a funding priority (Nelkin 2001), and negative expectations, by referring to avian flu as a natural bioterrorist in order to convince governments to stockpile the anti-viral drug Tamiflu as part of a pandemic plan (Nerlich and Halliday 2007).

However, when metaphors that have become "closed" in scientific discourse are "opened up" in the public domain, they can take on a new life of their own (Nerlich, Dingwall and Clarke 2002). When issues of science and public policy are bound up with one another, misappropriation of such images can be dangerous. A pertinent example is DNA's theory-constitutive "code" metaphor whose folk interpretation understands DNA functioning as a "programme" or "instructions," emphasizing prediction and alluding to biological determinism—an idea that intermittently resurfaces and has been used by right-wing proponents to try and convince that welfare and education are wasted on those who have "bad" genes (Herrnstein and Murray 1994). If people's essence is held to be in their genes, why spend money on trying to change something that cannot be changed? On the other hand, the code metaphor can also suggest the reverse given that codes can be altered—genetic "engineering" moots the very idea of change through facilitating health benefits and various other "upgrades" by "correcting design flaws." The idea of a "genetic supermarket" (Nozick 1974), where traits for children could be freely selected by parents, associated with the proactionary principle of policy making emerges when the engineering frame is angled in this direction. Of course, this frame can be angled the other way to strengthen different views, fuelling the fears of those who adhere to the precautionary principle that we could be "playing God," "creating Frankensteins," or a "genetic underclass" (see Fuller, this volume).

A crucial part of the public debate between these perspectives on the future of bioengineering will play out in a struggle of metaphors. If the proactionary camp is to prevail, metaphors will be a vital tool in promoting their agenda to gain the public acceptance and influence the political opinion that is necessary to change laws on germline engineering. They will also need to convince it is for the many and not just the elites, which may be difficult given that scientists clearly have vested interests in supporting bioengineering—not only in terms of commercial interests, but also because of the power and authority they stand to gain as the experts of these technologies.

CONCLUSION

Hopefully, it will now be clearer that while metaphors are an indispensable aid for purposes of knowledge creation and illumination that function in meaningful ways, particularly with nonobservables, they can also come to be perceived as nonmetaphorical and are also capable of and sometimes used for simplifying complex information in misleading and harmful ways. Hence, awareness of them and of their modes of operation, whether their use is ideologically motivated or not, is important so as not to fall victim to their power of persuasion and to enable us to challenge ones that do not meet our needs by either substituting or supplementing them with alternative ones.

BIBLIOGRAPHY

Ball, Philip. "A Metaphor Too Far." *Nature News*, February 23, 2011. www.nature.com/news/2011/110223/full/news.2011.115.html.

Boyd, Richard. "Metaphor and Theory Change: What is 'Metaphor' a Metaphor For?" In *Metaphor and Thought*, 2nd ed., edited by A. Ortony, 481–532. Cambridge: Cambridge University Press, 1993.

Bucchi, Massimiano. *Science and the Media: Alternative Routes in Scientific Communication.* London: Routledge, 1998.

Charteris-Black, Jonathan. *Corpus Approaches to Critical Metaphor Analysis.* Basingstoke: Palgrave Macmillan, 2004.

Duffy, Thomas M. and Donald J. Cunningham. "Constructivism: Implications for the Design and Delivery of Instruction." In *Handbook of Research for Educational Communications and Technology*, edited by D. H. Jonassen, 170–98. New York: Simon Schuster Macmillan, 1996.

Eliasmith, Chris. "Moving Beyond Metaphors: Understanding the Mind for What It Is." *Journal of Philosophy* C, no. 10 (2003): 493–520.

Herrnstein, Richard J. and Charles Murray. *The Bell Curve: Intelligence and Class Structure in American Life*. New York: Free Press, 1994.

Knudsen, Susanne. "Scientific Metaphors Going Public." *Journal of Pragmatics* 35, no. 8 (2003): 1247–63.

Lakoff, George and Mark Johnson. *Metaphors We Live By*. Chicago: University of Chicago Press, 1980a.

Lakoff, George and Mark Johnson. "The Metaphorical Structure of the Human Conceptual System." *Cognitive Science* 4, no. 2 (1980b): 195–208.

Lakoff, George and Mark Johnson. *Philosophy in the Flesh: The Embodied Mind and its Challenge to Western Thought*. New York: Basic Books, 1999.

Nelkin, Dorothy. "Molecular Metaphors: The Gene in Popular Discourse." *Nature Review Genetics* 2, no. 7 (2001): 555–59.

Nerlich, Brigette, Robert Dingwall, and David D. Clarke. "The Book of Life: How the Completion of the Human Genome Project was Revealed to the public." *Health: An Interdisciplinary Journal for the Social Study of Health, Illness and Medicine* 6, no. 4 (2002): 1363–93.

Nerlich, Brigette and Christopher Halliday. "Avian Flu: The Creation of Expectations in the Interplay Between Science and the Media." *Sociology of Health and Illness* 29, no. 1 (2007): 46–65.

Nozick, Robert. *Anarchy, State, and, Utopia*. New York: Basic Books, 1974.

Pea, Roy D. "Practices of Distributed Intelligence and Designs for Education." In *Distributed Cognition: Psychological and Educational Considerations*, edited by Gavriel Salomon, 88–110. New York: Cambridge University Press, 1993.

Schuh, Kathy L. and Donald J. Cunningham. "Rhizome and the Mind: Describing the Metaphor." *Semiotica* 149, no. 1 (2004): 325–42.

Semino, Elena. *Metaphor in Discourse*. Cambridge: Cambridge University Press, 2008.

Sternberg, Robert J. *Metaphors of Mind: Conceptions of the Nature of Intelligence*. Cambridge: Cambridge University Press, 1990.

Thibodeau, P. H. and Lera Boroditsky. "Metaphors We Think With: The Role of Metaphor in Reasoning." *PLoS ONE* 6, no. 2 (2011): 1–11.

Yob, Iris M. "Thinking Constructively with Metaphors." *Studies in Philosophy and Education* 22 (2003): 127–38.

Chapter 8

Memetics versus Human Extension

Round Two

Gregory Sandstrom

> The meme is itself a dream, a piece of ideology, accepted not for its truth but for the illusory power that it confers on the one who conjures with it.
>
> —Roger Scruton (2013)

It's difficult to engage with a dialogue partner who doesn't yet know that his or her proposed idea is academically obsolete and continues with self-serving word play in the face of the community.[1] My contention here is that memetics as a "Dawkinsian-Universal-Darwinism" (DUD 1983) strategy is just such an obsolete idea; a defeated ideology and not a promising or productive science, philosophy, or theology/worldview theme. Or, to say it another way: to "change over time" is not necessarily to "evolve." There are thus examples of "things that don't evolve" that are worth studying both individually and collectively. Nevertheless, the grand evolutionary dream Richard Dawkins once had to contribute a theory of "memes" (1976) worthy of studies of culture and society may be examined regarding its aspirations and eventual downfall.

This chapter assesses the end of the "memetics" paradigm in social sciences and humanities (SSH), while also exploring the outline and possibilities of Human Extension (Sandstrom 2014, 2015) becoming an alternative in the quest for collective knowledge and group learning. The Social Epistemology Review and Reply Collective (SERRC) itself is taken as a flourishing example of Human Extension on a global scale. Our primary questions to ask here are: What does collective knowledge in society "extend" from and to and how can we participate in extending it?

ROUND TWO: MEMETICS ON THE DEFENSE

In the SERRC Collective Vision "Memetics vs. Human Extension: Round One" (2014b), I compared Dawkins's "extended phenotype" approach with the "extended mind" thesis (EMT, Chalmers and Clark, Menary, Theiner, et al.) and introduced the notion of Human Extension (McLuhan) as an alternative to memetics. "Memes" were portrayed as a scientific misnomer, an empty term invented by a biologist/ethologist with social epistemology apathy and social science envy. Dawkins was shown to be lacking a (new) sociological imagination, thus reducing human knowledge and existence to strictly material or physical causes rather than elevating them above the mundane. Memetics was seen as displaying materialistic imperialism mainly for atheists and antitheists based on outdated DUD.

To repeat the conclusion there, memetics is already a discredited "field" that is neither scholarly nor empirically rigorous (cf. Edmonds 2005). This is no longer a controversial claim, though a few scholars (Aunger, Blackmore, Dennett, Heylighen, Sterelny, et al.) continue to use the language of memetics, while even more still speak of "cultural evolution" with its "blind variation and selective retention" (Campbell 1960). This chapter aims to show that such "aimless" dispiriting talk about human beings as mere "survival machines" for self-replicating "memes" held in a DUD framework is sadly dehumanizing rather than hopefully human oriented or edifying.

In Round Two, we face the intellectual task of ending memetics along with cultural evolution as poor examples of cause/effect reasoning in SSH. Memetics represents a kind of ideological bubble ready to burst, similar to a science bubble (Pedersen and Hendricks 2013) that has lost its epistemic and financial support. This challenge to memetics and cultural evolution should be taken as constructively competitive, rather than as destructively combative. Indeed, the biggest difficulty with DUD-inspired cultural evolutionary thinking is how one might move beyond a paradigm that holds that nothing in principle can *possibly be* beyond it. To "evolve" beyond "evolution" is simply inconceivable for DUDs because universal evolutionism would then be effectively ended. Thus, to tell a cultural evolutionist that culture is more appropriately and best understood as a nonevolutionary phenomenon is to fundamentally test and provoke their most cherished beliefs.

At a recent conference on "cultural evolution," leading memetics proponent, academic philosopher, and organizer Daniel Dennett acknowledged what he called "terminological headaches," admitting that "cultural group selection," "Darwinian," and "memes" are "burdened with legacies of ideological conflict" (2014). He also claimed that Dawkins's term "meme" has been hijacked to mean "any cultural item that 'goes viral' on the Internet" (2014). At the same time, Dennett was nevertheless not ready and plainly unwilling to give up memetics just yet. “We might as well go on using the term ‘meme,’” he stated, “to refer to any relatively well-individuated culturally transmitted item that can serve as a building block or trackable element of culture *however it arrives on the scene*” (2014, my italics).

That Dennett doesn’t seem to realize that cultural transmission can be studied without invoking “evolution” is one thing. The avoidance of origins, compare “arrival” in this quotation, however, is both blatant and revealing of the term’s shortcomings. The supposed “trackability” of “memes” has likewise not proven helpful or effective in SSH, that is, to the concept’s main audience. This is largely because of what even Dennett openly recognizes; the theoretical utility of “memes” is questionable and even doubtful. Many people have thus rejected memetics after having found it conceptually fruitless and even dangerously misleading. Yet, Dennett’s personal choice to nevertheless simply “leave open” the theoretical definition of memes, ignore the criticisms, and keep using the term anyway smacks of anti-intellectualism and irresponsible communication. Such a position is not only unscholarly in avoiding nonevolutionary social and cultural theory, but also shows an unfortunate faithlessness in the collective judgment and will of public intellectuals and everyday thinkers to correctly reject memetics. Let us look more closely at the core problems.

THE UN-GENESIS OF MEMES

> The origin of species—Darwin’s problem—remains unsolved.
>
> —Scott Gilbert et al. (1996, 361)

That Darwin didn’t solve the mysterious “origin” of species/life problem is not surprising and surely forgivable; that others mistakenly use “evolution” as a late-modern creation story is still a regrettable feature of dialogue broadly between materialists and theists. Memetics lacks a cultural origin(s) story other than Darwinian evolution based on materialistic ideology. Dennett’s (and company’s) “cultural evolution” notion is thus notably quite unlike Henri Bergson’s “creative evolution” (1911), Pierre Teilhard de Chardin’s Alpha to Omega story of evolutionary creation (1955) or Robert Bellah’s religious human evolution (2011). Dawkins’s and Dennett’s memetics thus attempts to substitute mediocre evolution for special creation, while Bergson, Teilhard, and Bellah intentionally tried to bring the two together in collaborative science, philosophy, and theology/worldview discourse.

One way to decisively overcome memetics is to inquire about the source(s) or origin(s) of supposed “memes.” Cultural evolution theories suggest that memes will arise or "emerge" inevitably due to external environmental pressures. In this view, human beings do not intentionally "create" ideas or make innovations but rather are physical hosts for mental viruses or parasites, such as religion, advertisements, patents, or catchy song jingles. Evolution according to this view is thus an anti-Creation position. The supposition goes that memes simply "poof!" into existence, outside of our internal cognitive control and intellect. Yet we can nevertheless still incisively ask: What do memes themselves supposedly extend from? Very little of value has come from proponents of memetics in reply, with their priority focus on evolutionary processes instead of creative origins.

Dawkins's idea of memes as "cultural replicators" likewise fails to persuade as a legitimate contribution to SSH. If there is no actual "self" involved—not in a traditional SSH sense—then obviously memes cannot "self-replicate." They are self-less. The contention made here instead is that any pseudo-self-conceptualization based on naturalistic biology or materialistic physics cannot possibly serve as an adequate substitute for a "real" human self in SSH. Dawkins's and Dennett's materialistic version of "self" thus lacks a higher, more holistic notion of "person" or "personhood" (cf. Smith 2011) suitable for SSH. Thus, in the typical model of "gene-culture co-evolution" (Huxley, Cavalli-Sforza, Trivers, Wilson, Boyd, Richerson, Jablonka, Wilson, Mesoudi, Sanderson, Laland, Deacon, et al.), group learning and dynamic behavior are naturalistically automatic and/or functionally deterministic, rather than humanistically[2] creative and voluntaristic, thus disqualifying it as an SSH approach.

As Roger Scruton notes regarding the "origins" of memes, "Even if there are units of memetic information propagated from brain to brain, it is not these units that come before the mind in conscious thinking" (2013). The primacy of the mental, in SSH, suggests the need to rethink memetics and cultural change-over-time with which the field of social epistemology can help us. Such rethinking is found below in the notion of Human Extension.

One might wonder: Why even bother confronting memetics if it is already outdated and declining? The answer: because a grain of truth and hope seems possible in the midst of the cultural evolutionary facade. Dawkins's memetics was aiming to address the legitimate theme of collective human knowledge, shared ideas and memories, yet as an epidemiological rather than epistemological phenomenon (Fuller 2012). This view suggests we become "infected" by knowledge and ideas. Many people were/have been taken in by this possibility, a fact that is important to openly admit. Thus, if we can naturalistically study human thoughts and ideas, wouldn't this solidify SSH as legitimate fields? Dawkins gave people false hope that his materialistic approach to culture would ably perform the existential trick, with universal evolutionism as his guiding principle and purpose.

Dennett likewise stresses that "some important elements of human culture evolve by processes strongly analogous to genetic natural selection" (2014). But obvious limits exist to those analogies. Moreover, misanalogies persist in that human culture "changes-over-time" in a very different way—teleologically, intentionally, and intelligently—in contrast to biology and genetics. A higher type of lowercase "intelligent design" (Sandstrom 2014), not found in natural science but in SSH, thus seems appropriate, even while safely rejecting the Discovery Institute's neocreationist politico-educational USAmerican IDist baggage.

GENERATING AND EXTENDING KNOWLEDGE: INDIVIDUAL AND COLLECTIVE

Human Extension constitutes a more appropriate, accurate, and fruitfully pregnant term than memetics for SSH studies of innovation and knowledge creation, development, and transfer. It highlights the teleological free will, choices, and actions (i.e., proactivity) of individuals living in various societies and cultures. The key research questions are: How do we, you, and I extend in our environments, in our relationships, in society? How do we intentionally shape our communities, while we are simultaneously also shaped by them? The Human Extension paradigm draws for support and dialogue on the extended mind thesis (Chalmers, Clark and Theiner), the history of extension theory in mathematics (Grassmann), philosophy and theology/worldview (R. Descartes, H. More and A. Conway), media studies (McLuhan), and on notions of extensive and intensive thinking in economic sociology (Marx, Marshall, Feldmann, Boserup and Thiel).

For EMT, of crucial concern are the varieties of internalism versus externalism (Carter and Palermos 2014) as they involve cognitive extension in society. Likewise, distributed group cognition and group knowledge are also part of the research program, along with social extension in terms of knowledge, education, and informatics (Extended Knowledge Project 2014). Unlike DUD, collective knowledge in society is not just a question of "group selection" versus "individual selection" or of environmental factors. Likewise, one need not limit themselves to only mental, but can also explore bodily (and if one believes in it, spiritual) extension. In terms of societies and cultures, Human Extension thus takes a variety of forms, which can be studied by SSH scholars and measured in institutions, contracts, events, activities, and experiences.

The focus on collective/group knowledge, distributed cognition, and community learning speaks directly to the main theme of this book. Human Extension acknowledges that people are nourished and inspired by communities and groups; that we participate in shared memories, emotions, and experiences on the basis of collective living, in addition to our individual subsistence concerns. We can thus affirm here that knowledge extends from and to people, including ourselves, even if we cannot always understand how, when, or where. The alternative desire of those who support memetics and cultural evolution is to somehow "naturalize" (read disenchant) knowledge of extension, which is on the one hand scientifically tempting, but also a utopian exaggeration that need not define humanity's quest for self-understanding.

When comparing memetics and Human Extension, the benefits in exploring collective knowledge as well as individual contribution should be obvious. What Human Extension promotes is that we should actively seek out collective activities and thus share in new ideas, experiences, solutions, and questions in a communal scholarly environment. This is more appropriate than just accepting materialistic "memes" that are passed our way to control us or that are claimed to autonomously "jump around" from person to person deterministically.

Dropping the language of memes entirely, we propose an alternative SSH research program that asks: How can we investigate the human extensions of knowledge? What insights does studying both the origins and processes of knowledge creation and distribution offer social epistemology? And why shouldn't we allow nonmaterialistic approaches such as Human Extension into the social scientific study of humanity in the realm of culture?

Even in the field of media studies, where the term "meme" will likely persist for the foreseeable future as a "viral" Internet phenomenon there is a misleading reification of human disempowerment and choicelessness via memetics. It is as if digital viruses can self-possess people with their fatalistic attitude. Thinking in terms of Human Extension openly acknowledges that each person has a choice: to click or not to click? Our media and real-life actions thus extend from actual choices, not only from external (digital) environmental pressures. This shows how the responsibility of voluntarism takes human-oriented priority over naturalistic evolutionary determinism.

As agents in collective knowledge and society, we do humanity a great disservice by allowing Darwinian acid to be poured on individual and collective intention, on free will, on choices. Though we all experience pressures, stresses, and tensions in our lives in different ways, we still possess the liberty to refuse ideologies like memetics that would reduce our humanity to merely atomic physical units rather than elevating us to improved self and community understanding.

CONCLUSION

To the primary question: What does collective knowledge in society "extend" from and to? We can answer simply: generative human choices and actions.

As ideologies, (neo-)Darwinism and memetics have both failed at their tasks.[3] There are many more important features of human life and understanding than what comes from applying Darwin's ideas in society and culture. The Darwinian education industry has thus unfortunately blocked collective progress in the electronic-information era, though it surprisingly still pretends to augment it. Let's instead now apply language more suitable to the current circumstances beyond the DUD paradigm.

One significant example of non-DUD views is the "visioneering" work of the SERRC (Cabrera, Leach, Navarro, Whitaker, and Orozco) and also the proactive thinking of Steve Fuller and Veronika Lipinska. Likewise, Human Extension can be explored to help us understand our meaningful existence on Earth given the intensions of environment and society. In our Collective Vision work, group epistemic agency (Kerr, Scalambrino, Martini, Reider and Olson) has been presented that epitomizes the strengths of SSH research adding to normative assessments of society and culture. In this sense, the SERRC is itself an example of Human Extension that spreads around the globe wherein the social epistemology program is being productively applied to each of these visions.

In the digital era, Human Extension is validated via "extension" cords, the Internet, mobile technologies, and so forth. We can even start to speak of human satellites as extensions of humankind (Sandstrom 2014a), looking to both our first and second lives exploring how we digitally or physically enhance and communicate among ourselves. The outdated Darwinian-Victorian view that would have us declare impotence to supersede biological evolution serves as a motivating factor to further explore Human Extension, even within a theistic or "spiritual" rather than materialistic framework.

Memetics now appears to us as a backward-looking ideology based on nineteenth- and mid-twentieth-century (neo-)Darwinism that is stained with "survival of the fittest" and "struggle for life" rhetoric. These notions are now constructively outcompeted for relevance as they have no sovereign life of their own outside of the people who ideologically embrace them. DUD is now fading away and only a desperate thinker would wish to reapply it.

Dawkins is not best interpreted as an "enemy of reason," a term he has condescendingly called Fuller. Rather, as a scientist he disenchants and defiles collective wisdom and fellowship with impunity. His memetic ideology will sink down into ignominy. As a nonmaterial or immaterial unit, information throws the materialistic DUD ideology into inevitable and irreversible disrepute. One need not criticize Dawkins personally; it is rather his materialistic worldview that disconnects him from properly anthropic collective learning in the electronic-information era.

It doesn't seem like we need go another round challenging memetics given the current circumstances. We can instead focus our attention and energy elsewhere. But by understanding what made it stay in the realm of potential ideas as long as it did, overcoming memetics can help us discover new possibilities that are more suitable for elevating SSH. The paradigm of Human Extension offers just such an explorative adventure in that personal and collective pursuit.

NOTES

1. "The meme is itself a dream, a piece of ideology, accepted not for its truth but for the illusory power that it confers on the one who conjures with it" (Roger Scruton 2013).

2. Steve Fuller speaks of the naturalistic and humanistic "ends of the social epistemology spectrum" (1987, 2).

3. "The 'track record' of Neo-Darwinism is parasitic on prior creationist breakthroughs over which Neo-Darwinists now claim sole ownership, and which creationists have yet to claim back as their own" (Steve Fuller 2005, 233).

BIBLIOGRAPHY

Bellah, Robert N. *Religion in Human Evolution.* Cambridge, MA: Harvard University Press, 2011.

Bergson, Henri. *Creative Evolution.* Translated by Arthur Mitchell. New York: Henry Holt and Company, 1911.

Campbell, Donald T. "Blind variation and selective retention in creative thought as in other knowledge processes." *Psychological Review* 67 (1960): 380–400.

Carter, Adam and Orestis Palermos. "Active Externalism and Epistemic Internalism." *Erkenntnis* 80, no. 4 (2015): 753–72.

Chalmers, David and Andy Clark. "The Extended Mind." *Analysis* 58, no. 1 (1998): 10–23.

Dawkins, Richard. "Universal Darwinism." In *Evolution From Molecules to Men*, edited by D. S. Bendall, 403–25. Cambridge: Cambridge University Press, 1983.

Dawkins, Richard. *The Selfish Gene.* Oxford: Oxford University Press, 1976, 1989.

Dennett, Daniel. "Perspectives on Cultural Evolution." SFI Cultural Evolution Workshop (2014). bit.ly/1oAlw7h.

Edmonds, Bruce. "The Revealed Poverty of the Gene-Meme Analogy—Why Memetics Per Se Has Failed to Produce Substantive Results." *Journal of Memetics* 9, no. 1 (2005). jom-emit.cfpm.org/2005/vol9/edmonds_b.html.

Extended Knowledge Project (2014). www.extended-knowledge.ppls.ed.ac.uk/.

Fuller, Steve. "Social Epistemology: A Statement of Purpose." *Social Epistemology* 1, no. 1 (1987): 1–4.

Fuller, Steve. *The New Sociological Imagination.* London: Sage, 2005.

Fuller, Steve. "Social Epistemology: A Quarter-Century Itinerary." *Social Epistemology* 26, nos. 3–4 (2012): 1–17.

Gilbert, Scott, John Opitz, and Rudolf Raff. "Resynthesizing Evolutionary and Developmental Biology." *Developmental Biology* 173, article no. 0032 (1996).
McLuhan, Marshall. *Understanding Media: The Extensions of Man.* Toronto: Signet, 1964.
Menary, Richard, ed. *The Extended Mind.* London: MIT Press, 2010.
Pedersen, David Budtz and Vincent F. Hendricks. "Science Bubbles." *Philosophy and Technology* 45, no. 3 (2013): 1–16.
Sandstrom, Gregory. *Human Extension: An Alternative to Evolutionism, Creationism and Intelligent Design.* Basingstoke: Palgrave Macmillan, 2014a.
Sandstrom, Gregory. "Human Satellites and Creative Extension." *Social Epistemology Review and Reply Collective* 3, no. 3 (2014b): 60–63.
Sandstrom, Gregory. *The Extension of Evolution.* Minsk: European Humanities University Press, forthcoming 2015.
Scruton, Roger. "Scientism in the Arts and Humanities." *The New Atlantis* 40 (2013): 33–46.
Smith, Christian. *What Is a Person? Rethinking Humanity, Social Life, and the Moral Good from the Person Up.* Chicago: University of Chicago Press, 2011.
Teilhard de Chardin, Pierre. *The Phenomenon of Man.* London: Harper, 1955.
Theiner, Georg. *Res Cogitans Extensa: A Philosophical Defense of the Extended Mind Thesis.* Frankfurt: Peter Lang, 2011.

Chapter 9

A "Dialectical Moment"

Desire and the Commodity of Knowledge

Patrick J. Reider

This chapter outlines a simplified conception of G. W. F. Hegel's account of history. Its purpose is to broadly frame a historical approach to social epistemology in which I argue society, knowledge, history, and desire are fundamentally interconnected. This historical approach denies the assumption that there is a fixed ahistorical account of knowledge and that there exists one best description of existence in which opposing accounts are necessarily false. I argue further that it operates as a productive method to address Steve Fuller's challenge: social epistemologists ought to provide ethical solutions to contemporary problems. In this regard, I make the case that a focused study of the interplay between history and desire provides an effective means to address the current ethical implications of knowledge acquisition and dissemination.

KNOWLEDGE IS IRREDUCIBLY SOCIAL AND A PUBLIC COMMODITY

If one is actively engaged in the acquisition and production of knowledge, one typically needs a monetary supplement to engage in one's craft. One also needs a social network upon which the materials of one's craft are rendered accessible. It matters not if someone requires an education, laboratory, library, or Internet resources, as in each of these instances, significant time, labor, and money are allocated to their productive management and maintenance.

Since institutions like schools, laboratories, libraries, and Internet resources are "managed" and they involve significant capital to ensure their maintenance and daily operation, they are necessarily tasked with certain goals, that is, one cannot sustain and successfully manage an expensive and complex system without well-defined goals. Goals coordinate many participants and direct the consumption of a wide range of resources—money, labor, intellectual, and so forth. Wide-ranging investments of resources and participation ensure that there are disparate collections of individuals with distinct interests and backgrounds, who all hope to wrangle some perceived good from their involvement in knowledge production.

If it is the case that the questions being researched require the supporting allotment of resources from managed (i.e., goal directed) instructions, it is also the case that the culture in which these spheres of interest operate are subject to norms that shape one's perception of an acceptable answer. Put differently, people tend to refrain from investing significant resources into projects with no accepted criterion of return: if I do X, I can expect certain benefits or privileges for others or myself. This requires some generally accepted criterion for epistemic achievement, which shapes one's (or the groups') expectation of what is required to obtain rewards for contributing or investing in it.

When knowledge production and acquisition have no clear sense of benefit (e.g., material, spiritual, physical, ethical, or privilege), the extent to which a community is willing to invest in it diminishes. This implies that, without some notion as to what others will require as evidence of epistemic success, one is unlikely to invest time or money in its acquisition and production. On these grounds, one can make a case that knowledge production and acquisition undergo the same social-economic forces as supply and demand of material goods.

If the above is true, the perceived good of knowledge acquisition and production cannot be a pure expression of a solitary individual but, to a varying extent, reflects diverse individuals with cooperating values. Even if I am a purely selfish being, my pursuit of knowledge acquisition and production will likely benefit me, as it concerns opportunities or privileges others are willing to confer on me for my participation in the epistemic process. The management of resources that make knowledge acquisition possible is also directed by a similar *telos* (i.e., the purpose that directs its application), in that knowledge is a commodity with social value, which has a certain role to play in the economy (in both the literal and the figurative sense) of the society that maintains its proliferation.

The concrete roles individuals and collectives play in shaping the questions asked in a society, along with the allotment of the resources required to enable their resolution, ought to be clarified and studied, because knowledge acquisition and production are public commodities and community-based resources. Here, the persistent struggle between individualists and egalitarians spills over into the various modes of knowledge production. Should private individuals, corporations, representatives of the people, or populist opinions shape and direct the course of knowledge production? These are issues that ought to equally concern the libertarian and socialist, the capitalist, and anticapitalist.

THE DIALECTICS OF KNOWLEDGE AND DESIRE

If it is the case that (1) knowledge is a deeply social phenomena that requires highly organized and cooperative social structures, (2) the individuals who willingly participate and maintain them, do so for some perceived good, and (3) knowledge production and acquisition consequently exist as a commodity, it will also be the case that the social phenomenon of knowledge is spurred on and shaped by desire. This is the case, as I will argue, because commodities, as a means to gain access to something else, are goal driven, that is, there is some desired outcome to be reached.

When the conditions of a particular culture permit individuals or portions of its population to yearn for something not existing (or not readily accessible to its members), strategies for their attainment inevitably create the conditions for new methods of investigation. In other words, old methods are employed in new ways, or they are rejected for new methods that promise to fulfill new goals. When changes in methodologies occur, there is a corresponding growth of how knowledge can be obtained and employed.

With each substantially different methodology, employment, and acquisition of knowledge, new horizons of possibilities are made available. Such developments can cause surprising ethical problems. For example, computer science has given rise to the unintended consequence of growing populations who actively labor to improve their online reputations, but fail to show the desire to cultivate improved means of interfacing with their face-to-face community. Such existential changes (i.e., changes in what is possible) generate the conditions for unexpected and unforeseen desires to take place. In turn, desires are greatly affected by the type of conduct and goods a particular culture endorses, makes available, or even brings to one's attention by making them taboo. Desire, like knowledge, is thus not merely a private enterprise, but a condition that can affect the modes of being or states of activity that transfix entire populations. For example, desires can possess our souls, in that they can (like knowledge) influence the internal principles that orient and shape one's life.

DESIRE AND THE COMMODITY OF KNOWLEDGE AS HISTORICAL FORCES

The historical trajectory of knowledge is largely directed by the questions we, not merely as individuals, ask, but far more importantly, the questions a given society is willing to invest its resources to discover. Such investments disclose the desires that underpin the impetus to invest in knowledge production and acquisition, when one determines what they are intended to achieve. Armed with such information, one can see the driving forces of the current forms of knowledge production and acquisition as stemming from a group's collective or cooperative desire to pursue some perceived good.

The above insight is obscured by the myth of pure reason, as propagated by rationalists and Kantians like Fichte, or the hard sciences, which often claim to work outside the realm of social influence. These orientations cover up the essential role society plays in the development of knowledge, because they fail to appreciate that desire, while not rational itself, directs and orients our rational goals. Consequently, they fail to note that the desires affecting our epistemic goals typically have a historical and cultural basis.

The popular "normative turn" in epistemology may at first blush appear to avoid these oversights. The normative turn refers to the attempt to understand "warrant" (what makes a claim acceptable and justifiable) by studying the complex network of expectations and accepted practices that underlie knowledge claims. Yet, a purely normative approach is also lacking, because it fails to make allotments for desire's effect on norms. Strong desires, once roused in individuals and populations, are rarely restrained by currently existing norms. They force societies to find new methodologies for obtaining the object of desire or rescinding its taboo status, which in turn results in new norms.

Understanding historical periods and the state of *eros* they create is no easy task. This is most clearly seen in the fact that history primarily concerns "events" as they shape, shake, break, and change societies. Events, however, are difficult to accurately portray, because they are made up of many individual occurrences that often lend themselves to numerous accounts concerning the manner in which they are interrelated.

Historical events have their origin in how a person feels entitled (as accepted by some segment of her culture) to pull together disparate occurrences as an "interpretation" of their "relatedness." Such narratives are always "interpretations," because the manner in which thoughts and actions affect the motivations, beliefs, and feelings of entire populations is not something that can be assessed by the chemist's litmus test (e.g., does the paper turn blue or pink?). Interpretations, as they are accepted or can be argued for within a normative framework, are the historian's ultimate court of appeals.

The interpretation an individual or society accepts is largely influenced by the manner in which a people want to see themselves and others, by what they think is relevant and important, and by what interests them; in short, by the things they desire—for interest, relevancy, or the importance of data cannot be shaped by facts until the narrative being told has set for itself a goal.

Goals are products of human desire, not the objective world of mechanisms and causal relations. While facts are undeniably important to historians, facts alone are equally inadequate for one to communicate history. Hence, history is constructed from an admixture of individual and cultural attitudes and desires. Since history is not a long strand of facts laid bare by empirical study, histories run the risk of being fictions.

There is reciprocity between history, knowers, and desire, because desire sets out goals that often establish opportunities for new methodologies. Likewise, new methodologies generate new knowledge and histories. In turn, new methodologies, histories, and knowledge create opportunities and products (material, cultural, and spiritual) that make fertile ground for new and unexpected desires.

The reciprocal interplay between desire and knowledge production represents, in part, one aspect of Hegel's conception of the dialectical movement of human reason in history. It also plays an important role in his conception of individual actors. This is especially the case in his account of "unhappy consciousness," a state which desires but fails to achieve an authentic way of integrating universal traits and principles (of one's culture) into one's individuality.

Substantial shifts in the phase of a people's art, religion, or knowledge are heralded by the new horizons of possibilities that enable cultures to change and grow in relation to their previous states. Such historical periods that lie between significant changes in a given culture can be seen as a "moment." Here a "moment" does not reference a mere occurrence in space time, but an entire way of acting and knowing as embodied and experienced by a specific population unified via a shared culture. In these regards, I interpret Hegel as being an early social epistemologist.

DESIRE, HISTORY, AND CONCEPTUAL RELATIVISM

Since the dawn of academic sociology in the late eighteenth century, there has been an increasing awareness that scientific, historical, religious, and aesthetic claims are in constant flux as they are replaced by the new or significantly altered claims of succeeding generations. This occurrence is not always, or even often, reducible to the existence of better data, which informs better inferences. Rather, such overturning of claims often foreshadows significant transitions in the goals and available choices present to individuals and populations.

When goals/desires shift, available research methods change. As alluded to above, this will affect the manner in which knowledge and truth is understood, because the norms (i.e., what is deemed acceptable, and the unspoken practices that are expected from other participants in the same social group) change when a population's goals and desires shift.

Similarly, different histories and conceptual accounts permit for alternative ways to describe existence. Varying descriptions do not entail that one account is correct and all other diverging accounts are wrong, because the manner in which existence is effectively described is not necessarily limited to one account. For instance, effective descriptions rest on interpretations of how best to meet the goals that circumscribe available concepts. Alternative concepts may provide numerous and useful ways to talk and think about existences, and their wrongness or correctness is rarely reducible to some demonstrable "fact" of the matter but rather how successfully the description provided achieves some goal in relation to the entities and events it characterizes.

The above conditions provide the opportunity for several competing and coherent systems of concepts to produce opposing claims concerning the same phenomenon. So while some descriptions may be unequivocally false, opposing descriptions do not entail that there can only be one true account. Nor should the affirmation of this recurring social phenomenon be misconstrued as an excuse to incorrectly claim that all differing views are somehow equal or deserve the same merit. Hillary Putnam developed such a nuanced understanding of concepts, which he called "conceptual relativism."

Hegel offers sophisticated analyses of history and its effects on the manner in which populations perceive the world, themselves, and others, and consequently, how we form knowledge claims. His views hold important parallels to "conceptual relativism." For Hegel, individuals, as well as entire populations, are constantly synthesizing new and more coherent ways of perceiving the world. Though Hegel generally believes the historical development of knowledge indicates a progressive improvement, he is also open to the insight that all previous views are not wholly false. Similarly, he does not claim that previous historical modes of knowledge, which contradict our current modes, are always antiquated or inferior to present views. This outlook toward knowledge is akin to conceptual relativism.

What is the root cause that renders such synthesizing to be an observable and recurring theme throughout human history? Well Hegel, like Kant, argues that reason itself *desires* unity. Simply put, rationality itself is motivated by desire—the desire for unified/coherent accounts.

Before moving on, I wish to drive home the point that it is a myth to believe that newer knowledge claims are always governed by better data or improved deduction or induction. Knowledge is largely the achievement of clearly defined goals, accepted means of achieving these goals, and the successful completion of them. In such cases, the manner in which existence is interpreted, insofar as it is deemed to have been given an acceptable characterization via concepts, is goal driven by the desires it serves. And since there are many different desires that can be met and successfully achieved in a variety of ways, there exists dynamic and often opposing descriptions of the same phenomenon.

The above holds serious epistemic repercussions. For instance, the epistemic merit of one class of claims over another, when they meet their own criteria and achievement, tends to be reducible to a matter of *choice*, as it concerns what a person or society desires. Of course, there are many systems that fall apart from the incoherency of their own standards or the unreliableness or un-realizable-ness of their goals. Similarly, there are others that simply fail, because that which it describes cannot be observed (when it states the existence of observable traits) or its inferences fail to pan out. Consequently, truth is nuanced, in that it not only permits accurate descriptions in numerous ways, but also permits falseness in unlimited enumerations.

PLATO, FULLER, AND DESIRE

Where then is the work of the social epistemologist, if he or she wishes to do more than merely chronical the different developments of human thought in their various modes or expose their faults? Fuller (2014) offers a compelling challenge:

> The job of the social epistemologist—very much in the spirit of Hegel—is to discern the vehicles of these concepts at particular moments that have the capacity to expand the horizons of the human condition. . . . Thus, I have been drawn to Hilary Putnam's (1978) "pessimistic meta-induction" about the history of science; namely, that we have little reason to think that the agreed general theories of today are the ones most likely to be leading inquiry a century from now, just based on past predictions of this sort. For me this is the true take-home message of Kuhn's theory of scientific change, if we consider history "philosophically"—which is to say, *not* inductively: in other words, to learn from the past in its totality, not simply from a finite range of previous moments. The background intuition to this line of thought (which I see in Popper) is that any sort of consensus should be diagnosed rather than built upon—that is, regarded as a symptom of something missing, given that humans are fallible beings who seek safety in numbers. (33)

What should we do with the new horizons achieved from the "spirit of Hegel's" conception of history? Fuller argues that social epistemologists "should challenge people to embrace an ever expanding existential horizon, which implies openness to change—but not mere adaptation: Once one discovers the curve, one should try to get ahead of it" (34).

One possible response to this challenge has already been provided by Hegel's conception of "Unhappy Consciousness." As Hegel suggests, it is natural for people to want to belong and be connected to their communities, for this is how they discover themselves. Yet, at the same time, people also desire to act upon their unique individuality, which leads to internal and external struggle. The resolution of this struggle begins when individuals become aware of what they can make and do (i.e., their existential horizon) within the given framework of their social situated-ness (e.g., the worker's relation to his boss, i.e., master/slave dialectic). One way to interpret Hegel is to read him as claiming that individuals achieve authentic freedom by finding realizable ways to influence one's social conditions, while at the same time, allowing one's individuality to flourish as it undertakes its social responsibilities and faces up to social realities.

How then should we proceed with this Hegelian framework, while pursuing Fuller's challenge? This question is especially vexing when we consider that an appreciation for history and conceptual relativism does not provide any hard and fast metric by which to evaluate our current circumstance, direct our actions, or resolve our ethical dilemmas.

Let us consider what we have covered so far to glean an answer. Knowledge is always a social phenomenon, it is necessarily situated in a particular society or social group, and it historically develops. In each of these instances, desire is driving the current expression of knowledge and historical interpretation. Desire is thus critical for not merely understanding knowledge and history, but also for understanding how one can cope with new epistemic developments and the moral possibilities they engender. In this regard, I affirm with Plato that we ought to pay careful attention to our private and culturally driven desires. Plato warns individuals and cultures that promote ceaseless expansion of desire, especially when they become contradictory, will sting individuals and entire populations into madness, as if each new passion were a poisonous viper striking one's soul.

If the above is true, our epistemic endeavors ought to be yoked to the following considerations. What desires have we created as we continue to pursue knowledge, not for the love of wisdom, but rather for its use as a commodity? What new desires is this commodity-driven model of knowledge in the process of creating? Do these desires interfere with the productive management of our lives? And finally, who do such desires serve, in that one may well possess desires that do not properly serve oneself, but others?

I have argued that the above views outline the areas of study required to successfully navigate the minefield of desires our society, knowledge, and history have set before us as an individual, culture, and nation.

BIBLIOGRAPHY

Fuller, Steve. "Social Epistemology: The Future of an Unfulfilled Promise." *Social Epistemology Review and Reply Collective* 3, no. 7 (2014): 29–37.

Hegel, Georg Wilhelm Friedrich. *Lectures on the Philosophy of World History: Introduction, Reason in History*. Translated by H. B. Nisbet. New York, NY: Cambridge University Press, 1975.

Hegel, Georg Wilhelm Friedrich. *Phenomenology of Spirit.* Translated by A. V. Miller. Oxford, UK: Oxford University Press, 1977.

Putnam, Hilary. *Meaning and the Moral Sciences.* London: Routledge & Kegan Paul, 1978.

Putnam, Hilary. *Reason, Truth and History.* Cambridge: Cambridge University Press, 1983.

Chapter 10

Navigating the Dialectics of Objectivity

Guy Axtell

"Objectivity" is a theoretical concept with diverse applications in our collective practices of inquiry. It is also a concept attended in recent decades by vigorous debate, including but not restricted to the contributions of scientists and philosophers. In part because the authority of science is supposed to result from its objective methods and procedures, objectivity has been described as an "essentially contested," and even an "embattled" concept. That objectivity should have become the troubling and controversial concept that it is today was probably inevitable given the ferociousness of certain debates among academics since the breakdown of the positivist paradigm.

Recent movements in the theory of knowledge, most especially social and feminist epistemology, have been a source of both criticism of standard accounts of scientific reasoning associated with positivism, and of "reconstructive" accounts of objectivity. But many early critics of the value of the concept have in recent decades agreed with Helen Longino's stance that the concept should be "reconstructed" to fit contemporary views, rather than rejected. The standard account made the roles that values play in scientific practice especially controversial. Contemporary views still maintain a distinction between epistemic and social values, but, following Longino's insistence that we must deconstruct the rational/social dichotomy, do not necessarily view the role of social values as threatening epistemological relativism—an assumption that positivists and their radical historicist critics apparently shared. Longino's feminism "understands the cognitive processes of scientific inquiry not as opposed to the social but as themselves social" (1993, 260), which means taking social values positively, as aiding rather than necessarily or always detracting from objectivity.

There are many ways to understand the relationship between cognitive and social values, and the effects they have of scientific practice. Elsewhere I develop resemblances and differences between those I call *entanglers* and *disentanglers* (Axtell 2015, forthcoming). This distinction is *not* closely allayed with that between science and technology studies (STS) and philosophy of science, and that may be to its advantage. What proponents of these models share is that they view scientists' interactions as a source of both scientific error *and* scientific objectivity. Machamer and Osbeck (2004) say that all learning takes place within the *socio-cognitive*: "The social is built into the very conception and application of objectivity, and hence into the scientific ideal and cannons of practice," such that it is ultimately "impossible to determine where the social begins and the cognitive ends" (85, 88). Entanglers want us to acknowledge and sometimes even to encourage overlaps and interconnections between cognitive and social values in scientific practice. Impurities, interactions, ecological adjustments, and of course entanglements of facts and values, and of different kinds of values, are ideas associated with entanglers.

Projects of disentanglement are just as important for some feminists as demonstrating entanglements and epistemic impurities are for others. Disentanglers may find such entanglement descriptively accurate, but highlight normative ways of parsing values with the goal of improving science policy deliberation. Disentanglement represents attempts of all kinds to fine-tune the scientific traffic with values through ways of sorting values. The proposed ways of sorting are numerous, including sorting by stage within the scientific process, the social role of the agent, and so forth. Disentanglers urge a clearer differentiation between descriptive and normative claims in the debate over science and values. Anderson insists that the entanglement of fact and value in scientific research

> does not mean that factual and value judgments play the same roles in inquiry, much less that all claimed facts are political or ideological. Value judgments guide inquiry toward the concepts, tools, and procedures it needs to answer our value-laden questions. But facts—evidence—tell us which answers are more likely to be true. (2004, 19)

DIALECTICS AS THE PULSE OF PROGRESS

A dialectics of objectivity suggests a negotiative conception of the means and ends of inquiry, one with clear reference to the active, adjectival sense of objective *inquirers*. In *Rethinking Objectivity* (1994) Megill outlines four competing senses of objectivity: the absolute, the disciplinary, the dialectical, and the procedural:

> A striking feature of both absolute and disciplinary conceptions of objectivity is their negative relation to subjectivity. Absolute objectivity seeks to exclude

> subjectivity; disciplinary objectivity seeks to contain it. . . . Phrases like "aperspectival objectivity" and "view from nowhere" really draw attention to . . . negativity. In contrast, dialectical objectivity involves a positive attitude toward subjectivity. The defining feature of dialectical objectivity is the claim that subjectivity is indispensable to constitute the objects. Associated with this feature is a preference for "doing" over "viewing." . . . In other words, *subjectivity* is needed for *objectivity*; or, as Nietzsche put it, "objectivity is required, but is a positive quality." (8)

The special emphasis of a dialectical approach is on how something is constituted as an object for inquiry through interplay between researchers and that which they study. Objectivity, proponents of a dialectical model hold, is not a property of a right method or a steady state, but rather describes a *process* carried out actively through communicative interaction and comparison. But we need to pay special attention to how the content and value of the ideal of objectivity differs across scientific disciplines. Some philosophers of science connect this to metascientific pluralism, pointing out that different conceptions of aims and methods, and associated disciplinary norms of objectivity differ not only across the divide between natural and human sciences, but within each basic kind of science as well. Epistemic norms are dynamic, historical, and contextual. What feminist epistemologists like Heather Douglas refer to as the "irreducible complexity" of objectivity also embodies a substantial pluralism about objectivity's "modes" and their relevance to different scientific or nonscientific studies.

Steve Fuller refers to dialectics as "the pulse of scientific progress" (2004, ch. 5). He points out that criticism is productive only under certain conditions. The crucial test or "quick kill" is shown to be unlikely by historicists and moderate confirmation holists, but criticism of a scientific theory takes new and potentially more productive forms when applied to a research program seen as extending over a substantial period of time. For example, Fuller discusses Lakatos's (1976) idea that by reinterpreting a counterexample to a scientific theory as a boundary condition for future versions of the theory, "error elimination is made into a genuine collective learning experience, whereby a *prima facie* negative episode in the theory's history becomes a feature of its logical structure" (35). Such adjustments to theories over time are better seen as dialectical rather than strictly deductive, something which logicist science and constructive empiricism (which focus primarily on the *synchronic* relationship of empirical fit) will likely overlook. All of the sciences face periods of local underdetermination of theory choice by available data and methodological rules. But the eventual "dialectical resolution" of these situations shows again the lesson that Duhem first brought to light, that "theory choice is rarely, if ever, *forced* upon scientists" (37). This view is associated with a moderate historicism about norms of objectivity.

OBJECTIVITY AND THE REPLICATION CRISIS IN SOCIAL PSYCHOLOGY

One current example of the dialectics of objectivity that deserves attention by social epistemologists is the controversy surrounding p-hacking and the "replication crisis" in social psychology.[1] Social scientists have recently been making quite strong claims about others being engaged in so-called p-hacking. The claims range from intentional scientific fraud to regrettable scientific error, to complete exoneration. The "navigation" of this debate involves a number of questions that psychologists and social epistemologists need to ask. Some of these questions are:

- Are biases evident among working scientists within the lab that encourage the coding of a behavior in a way that supports their hypothesis?
- How does the discussion of replication problems illuminate the contextuality of scientific methods?
- Why have journals in social psychology previously largely ignored studies that attempt to replicate previously performed experiments, and has this contributed to the current "crisis"?
- How important or unimportant are replications (following a previous study's exact methods to recheck their data) for maintaining and improving the objectivity of research practices in this field, or in others?
- Are there other protocols that could effectively monitor and correct for bias in how behaviors are coded?
- Are fields like economics as prone to these problems as psychology is? How widespread is so-called p-hacking, and how large an incentive for it is personal advancement and pressure to publish?
- What should we say is actually shown by the failure of replication (failure to confirm the data and conclusions drawn from the original study)? What degree of opprobrium (sharp criticism and censure), if any, is appropriate when one's finding is not replicated when others try to perform those same experiments?

UNDERDETERMINATION PROBLEMS IN SCIENCE

Metascientific pluralism reflects differences in the objects, aims, and methods of the various sciences. My version of it uses the frequency with which a field of study is best by local underdetermination problems, as an indicator of the types of theory virtues or cognitive values practitioners rely upon as important to theory choice. How characteristic it is for a discipline to operate under conditions of deep underdetermination tells us much about the balance of theory virtues and personal virtues—what Pierre Duhem (1991) would call *bon sens—that function* as criteria of adequacy in that field. But notice that this use of the depth and frequency of underdetermination problems to make comparisons among the sciences does not mirror, but rather cuts across the usual distinction between natural and social, or again "hard" and "soft" sciences. Both historiography and theoretical physics, for example, for all their other differences (typically debated according to contrasts like those between *verstehen* and nomological explanation), are chronically subject to underdetermination-of-theory worries.

A related thesis is the potential synergies between moderate historicism about epistemic norms and virtue-theoretic approaches in epistemology. Considerations stemming from underdetermination problems motivate the claim that historicism requires agent-focused rather than merely belief-focused epistemology; embracing this point helps historicists avoid the charge of relativism. Considerations stemming from the genealogy of epistemic virtue concepts motivate the claim that character epistemologies are strengthened by moderate historicism about the epistemic virtues and values at work in communities of inquiry; embracing this point helps character epistemologists avoid the charge of objectivism (see Axtell 2012).

THE NORMATIVE PROJECTS OF SOCIAL, FEMINIST, AND VIRTUE EPISTEMOLOGIES

Social, feminist, and virtue epistemologies overlap in interesting ways, and their overlap is nowhere more clear than in the normative projects that each supports. For Fuller, the "epistemological" in social epistemology refers essentially to its normative projects, reflecting Fuller's basic assumption to treat epistemology as a normative enterprise (1988, 3). To the extent that Fuller's sometimes sharp critiques of "analytic social epistemology" are a reflection of his own commitment to sociological eliminativism, they lack much resonance with philosophers. It is difficult to make naturalism support fashionable reductionisms of this or any other sort. Harder still, to promote normative projects of improving the management of knowledge and science policy, while at the same time "eliminating" the contrast between the social and cognitive values by reducing the latter to the former. If one's view is that philosophers all rely on intuitions as a test of theory, or that "'epistemic injustice' as invoked by analytic social epistemologists looks like a high-minded way of entrenching academic norms of political correctness" (Fuller 2014, 33), one is arguably just protecting a vested conception of social epistemology rather than inviting genuine interdisciplinary dialogue.

Social, feminist, and virtue epistemologies all share Fuller's prescriptive project of *improving* scientific and policy-making-related practices, but they can do so without taking a specific stance on metaphysical issues dividing scientific realists, empiricists, and social constructivists. Social epistemologists might have been better served by following Helen Longino's prescription to let such reductionisms die along with the rational/social dichotomy. The conceptions of social epistemology proposed across STS and philosophy may prove complementary or they may not, but it does little good to polarize them by exaggerating their differences. Relatedly, it is always a temptation to overstate the expansion of epistemology by describing a new project as *supplanting* rather than *supplementing* traditional concerns. Character epistemologies too often did this themselves in their early days.

Despite these concerns about the "two camps" account of the future of social epistemology, a good deal of contemporary work—including this volume in particular and Reider (ed. 2015, forthcoming)—invites a substantially more progressive, and less polemical, discussion of social epistemology.[2] These more recent exchanges I think demonstrate that social, feminist, and virtue epistemology share strongly prescriptive projects, while each bringing their own formidable resources. Feminist, social, and virtue epistemologies have been especially concerned with the reception of testimony, and so with issues of trust, credibility, and dependence upon others for most of what we think we know. Different schools of thought concerned with social epistemology have also been converging on the "decentered" conception of cognitive agency argued for by Patrick Reider (ed. 2015, forthcoming), among others. Individualistically orientated epistemology (especially apparent in internalist thought) values epistemic self-reliance or autonomy: direct knowledge is best; dependence on the testimony and the trustworthiness of others is of lesser value. Social epistemology redresses this traditional bias toward independently acquired versus dependently acquired knowledge. If knowing is taken more as a collective enterprise, then our epistemic dependence upon others is significantly greater.

To the extent that knowledge relies on testimony and trust, not only truth seeking, but also ignorance can play a positive role. So attempts to integrate feminist, social, and virtue epistemologies include Miranda Fricker's voluminous work on epistemic injustice and Cynthia Townley's (2011) and Phil Olson's more specific research (this volume) on "oppositional ignorance," which involves the interested, invested, or active refusal to bear certain epistemic burdens. Such work highlights the philosophical upshot of *epistemic dependence*, while drawing attention, on a more practical level, to normative questions of epistemic justice and injustice in the social distribution of epistemic burdens.[3] At the intersections there is little reason to dichotomize between character epistemology's interest in person-level virtues of social justice and the social epistemologist's interest in what social and institutional settings best nourish conditions of epistemic justice.[4] We can easily employ a normative language of group and institutional virtues and vices so long as these concepts do not presuppose mysterious entities like group minds. The robustness of Fricker's social epistemology is recognizable in her insistence upon a "nonsummative" conception of groups.

> This is a crucially important point not only from the perspective of the philosophy of collectives but also from the first-personal or ethical-political point of view of identifying those collective practices of racism and other kinds of discrimination or prejudice that are not a matter of individuals' personal attitudes. (Fricker 2009)

ACTIVE EXTERNALISM: EPISTEMIC AGENCY DECENTERED

Extended, collective, and distributed cognition each point to different forms of knowledge-conducive epistemic dependency, and are studied through different connections with the sciences (see Duncan Pritchard 2014). Recent advances within philosophy of mind and cognitive science have thrown light on extended cognition, the many ways in which human cognition is enhanced or enabled by instruments, human enhancement, and the experience of virtual reality. Different schools of thought concerned with social epistemology have also been converging on a distributed conception of cognitive agency. The idea of distributed knowledge considers the importance of social divisions of labor and dependency relations. Cognition is distributed among these suprapersonal entities too. Knowledge-producing or knowledge-maintaining institutions such as universities, research laboratories, and libraries, along with our general dependence upon the testimony of other persons and collectives such as news sources we trust, are just some of the ways that dependency relations structure our epistemic agency. Collective epistemology, for which Margaret Gilbert gives Emile Durkheim some credit to as a source, studies collective knowledge, collective belief, and collective responsibility. Group deliberation and the role of deliberative virtues have also received more attention in recent years. A title Palermos and Pritchard (2013) give to these interests is active externalism, which can be partly defined in terms of the denial of cognitive individualism or the idea that "human cognition is not always limited to our biological capacities" (109).

So to summarize, while a kind of methodological individualism characterized many projects undertaken by analytical philosophers, analytical philosophy itself is far too broad to be characterized by such an assumption. What we *can* say is that extended, collective, and distributed cognition are all drawing keen interest among epistemologists today, but the issues they raise were largely ignored in traditional normative epistemology committed to an individualistic conception tied to an ideal of the independent, self-reliant, and unbounded knower.

NATURALIZING AND SOCIALIZING EPISTEMIC VIRTUES

Scientific pluralism has been presented as an alternative to both the "unity of method" that Hempelian logical empiricists demanded, and to the epistemological relativism of some of the post-positivists. This middle path is arguably facilitated by a naturalized epistemology of habits or virtues. So let me conclude with three brief connections between virtue-theory in the philosophy of science, and the shared normative concerns of social epistemologists. A connection between pluralism and epistemic virtues is suggested by the editors of a notable collection, *Scientific Pluralism* (Kellert et al. 2006) when they explain,

> Philosophers of science have begun to advance pluralism at the metascientific level, most notably with respect to epistemic virtues. A variety of views regarding the role, status, and identity of scientific or epistemic virtues has been advanced in the philosophical literature. . . . [Some pluralists claim] that which virtues should hold what degree of regulative status in any given research project is a function of features specific to the problem and of the particular aims of the research. (Introduction, p. x)

Philosophers of science such as John Dupre make a second connection in suggesting that

> the successor to the quest for demarcation criteria between science and non-science may be an account of theory virtues that characterize scientific reasoning. . . . We are much better off to think in terms of epistemic virtues, features of an investigative practice that confer credibility. No doubt the cardinal empirical virtue is a proper connection with empirical evidence, which is the large grain of truth in the criterion of falsificationism. (2010)

The implication is to abandon the hierarchy of the sciences ideal, but without invoking any rigid egalitarianism either:

> We could try to replace the kind of epistemology that unites pure descriptivism and scientistic apologetics with something more like a virtue epistemology. . . . Many plausible epistemic virtues will be exemplified as much by practices not traditionally included within science as by paradigmatic scientific disciplines. . . . No sharp distinction between science and lesser forms of knowledge production can survive this re-conception of epistemic merit. It might fairly be said, if paradoxically, that with the disunity of science comes a kind of unity of knowledge. (Dupre 1993, 242–43)

Finally, "naturalized" virtue epistemologies for the philosophy of science eschew the intuitionism of the tradition of rational reconstructionism in favor of an empirical study of learning how we learn (see, e.g., Fairweather 2014). The "separate culture" notion associated with philosophy's autonomy on the one side, and with intuitionistic methods on the other, is dying fast. The goal of naturalized epistemologies is not to justify the choices of past scientists as rational, but a "bootstrapping" project of piecemeal improvement that depends upon the recognition of the bounded or ecological nature of our epistemic agency. To this end philosophy draws upon the special sciences for a contextualized mapping of which methods promote which field or context-specific cognitive aims. Sources of error and failures of objectivity are just as important to study as are successes. In this way as well, attempts to articulate a more virtue-theoretic philosophy of science highlight the facts of our epistemic dependence, but also strongly overlap with some of the prescriptive or guidance-related projects of social epistemology.

NOTES

1. A good place to start is the debate between Kurzban, Kahneman, and others at edge.org/panel/robert-kurzban-p-hacking-and-the-replication-crisis-headcon-13-part-iv.
2. Finn Collin, 2013, makes a good point in writing that "Analytic social epistemology's methodology conforms closely to what Popper advocates under the label of 'piecemeal social engineering,' while the strategy recommended by Fuller verges towards 'utopian social engineering' which is the project of effecting large-scale and radical transformations of society." A good test case of this may be positions on Humanity 2.0, since Fuller's position strikes me as reflecting many similarities to Kurzweil's utopianism, even if Kurzweil and the majority of posthumanists are political libertarians who would scoff at Fuller's call for the radical democratization of technology's benefits.
3. See Fricker and Townley 2011. In virtue epistemologies, the expansion beyond the "project of analysis" of knowledge, and beyond veritist conceptions of our cognitive aims, is especially evident in "responsibilist" approaches. To review recent work visit *JanusBlog: The Virtue Theory Research* Forum at janusblog.squarespace.com.
4. See Reider (ed.) 2015, with contributions by Sanford Goldberg and Steve Fuller. For another account that is suspicious, like myself, of the contrast of the "two camps" of social epistemology, see Palermos and Pritchard 2013.

BIBLIOGRAPHY

Axtell, Guy. "The Dialectics of Objectivity." *Journal of the Philosophy of History* 6, no. 3 (2012): 339–68.

Axtell, Guy. *Objectivity*. Cambridge, UK: Polity Press, 2015 forthcoming.

Carter, J. Adam and Orestis Palermos. "Active Externalism and Epistemic Internalism." *Erkenntnis* 80, no. 4 (2014): 753–72.

Collin, Finn. "Two Kinds of Social Epistemology." *Social Epistemology Review and Reply Collective* 2, no. 8 (2013): 79–104.

Daston, Lorraine L. and Peter Galison. *Objectivity*. New York: Zone Books, 2007.

Duhem, Peter M. M. *The Aim and Structure of Physical Theory*. Translated by Philip P. Wiener. Introduction by Jules Vuillemin. Foreword by Louis de Broglie. Princeton, NJ: Princeton University Press, 1991.

Dupre, John. *The Disunity of Things*. Cambridge, MA: Harvard University Press, 1993.

Dupre, John. "The Disunity of Science." Interview with Paul Newall (2010). www.galilean-library.org/site/index.php/topic/3302-john-dupre-the-disunity-of-science/.

Fairweather, Abrol, ed. *Virtue Epistemology Naturalized*. New York: Springer, 2014.

Fricker, Miranda. *Epistemic Injustice*. Oxford: Oxford University Press, 2007.

Fricker, Miranda. "Can There be Institutional Virtues?" In *Oxford Studies in Epistemology: Volume 3*, edited by Tamar Szabó Gendler and John Hawthorne, 235–52. Oxford: Oxford University Press, 2010.

Fuller, Steve. *Kuhn vs. Popper*. New York: Columbia University Press, 2004.

Fuller, Steve. "Social Epistemology: The Future of an Unfulfilled Promise." *Social Epistemology Review and Reply Collective* 3, no. 7 (2014): 29–37.

Kellert, Stephen H., Helen E. Longino, and C. Kenneth Waters. *Scientific Pluralism*. Minneapolis: University of Minnesota Press, 2006.

Lakatos, Imre. *Proofs and Refutations: The Logic of Mathematical Discovery*. Cambridge, UK: Cambridge University Press, 1976.

Longino, Helen. "Essential Tensions—Phase Two: Feminist, Philosophical, and Social Studies of Science." In *A Mind of One's Own: Feminist Essays on Reason and Objectivity*, edited by Louise Anton and Charlotte Witt, 257–72. Boulder: Westview Press, 1993.

Machamer, Paul and Lisa Osbeck. "The Social in the Epistemic." In *Values, Science and Objectivity*, edited by Paul Machamer and Gereon Wolters, 78–89. Pittsburgh: University of Pittsburgh Press, 2004.

Megill, Allan, ed. *Rethinking Objectivity*. Durham, NC: Duke University Press, 1994.

Palermos, Orestis and Duncan Pritchard. "Extended Knowledge and Social Epistemology." *Social Epistemology Review and Reply Collective* 2, no. 8 (2013): 105–20.

Pritchard, Duncan. Project description for "Virtue Epistemology, Epistemic Dependence and Epistemic Humility" (2014). humility.slu.edu/team/duncan-pritchard/.

Reider, Patrick J. *Social Epistemology and Epistemic Agency: De-Centralizing Epistemic Agency*. London: Rowman & Littlefield International, 2015 forthcoming.

Townley, Cynthia. *A Defense of Ignorance: Its Value for Knowers and Roles in Feminist and Social Epistemologies*. Lanham, MD: Lexington Books, 2011.

Chapter 11

Epistemic Burdens and the Value of Ignorance

Philip R. Olson

The landscape of epistemology is changing. Modern western epistemology has largely focused on questions specifically having to do with knowledge: for example, whether it is possible to know anything; the relationship between knowledge and the state of the external world, or between knowledge and the mental states of the knower; and the grounds for rightfully asserting that one knows what one claims to know. Recently, however, many scholars have sought to expand the scope of epistemology to include subject matters other than knowledge and its justification (Axtell 2008). Often, this expansion takes the form of inquiries into lofty epistemic goods such as understanding and wisdom. But the expansion of epistemology has also embraced what might seem a more lowly subject matter: ignorance. Running against the grain of traditional epistemology, growing numbers of scholars have furthered a project that, write Shannon Sullivan and Nancy Tuana (2007), aims to "identify different forms of ignorance, examining how they are produced and sustained, and what role they play in knowledge practices" (1). Whether understood as a unified project or a set of related projects, the epistemology of ignorance has evolved over the years, as new scholars contribute fresh insights into the undeniably "complex phenomena of ignorance" (1). The scholars who have most ardently and skillfully advanced the epistemology of ignorance are those deeply engaged in feminist theory and race theory, for whom "the study of ignorance is a valuable tool for liberatory epistemologies" (Tuana and Sullivan 2006, vii–ix). Indeed, the study of ignorance has joined forces with studies of epistemic justice and injustice to identify and confront various knowledge practices that engender and sustain oppression.

In "Epistemologies of Ignorance: Three Types," Linda Alcoff sets Charles Mills's conception of ignorance apart from the conceptions of ignorance developed by Lorraine Code and Sandra Harding. Drawing attention to the situatedness of knowers, Code and Harding emphasize ignorance's status as a limitation or lack of knowledge. In contrast, Charles Mills illuminates a kind of ignorance that is no mere lack of knowledge, but which instead consists of positive epistemic practices that are used by the socially empowered to marginalize and subjugate whole groups of epistemic agents. This *substantive* ignorance is a feature of "structural social conditions" within which certain "identities and social locations and modes of belief . . . are in some cases epistemically disadvantaged or defective" (Alcoff 2007, 39–40). Ignorance of this sort wields a veiled power, concealing—even from itself—the habits of attention, perception, judgment, and memory by which it marginalizes and subjugates whole groups of epistemic agents. To confront this power is to confront a source of epistemic injustice, and to make possible the initiation of potential correctives. Indeed, recent epistemologies of ignorance have undertaken the critique of substantive, structural ignorance in the name of those who are oppressed by the powerfully ignorant.[1]

Whether ignorance is understood as a lack of knowledge, or as a substantive epistemic practice, ignorance is typically thought to be inferior to knowledge from an epistemic point of view. Yet Cynthia Townley (2006, 38) contests the "global devaluation of ignorance" while also insisting upon a limit to the value of knowledge.[2] The limited value of knowledge may come into focus through the study of a variety of familiar epistemic goods like understanding and wisdom, the values of which are not reducible to their contributions to the value of knowledge.[3] But Townley makes a compelling case for the more controversial claim that ignorance *itself* has value as an epistemic good—a good that is among those that "are not reducible to the value of increasing knowledge" (2011, x). Critical of the excessive and myopic drive toward knowledge acquisition that she calls *epistemophilia*, Townley argues that ignorance possesses intrinsic value as a necessary condition for empathy and trust, which in turn are essential for responsible participation in epistemic communities. More specifically, Townley argues that empathy and trust require what she calls "simple ignorance," that is, a kind of ignorance that entails both "lacking knowledge, and refraining from remedying that lack" (23). Despite her recognition of the value of simple ignorance, Townley joins Mills, Sullivan, and Tuana in their scorn of substantive ignorance. According to Townley, it is important to study substantive ignorance (which she calls invested or interested ignorance) only because such ignorance facilitates unjust distributions of trust, credibility, and labor, and is thus "a kind of ignorance that should be challenged or contested" (51, n. 10). Townley's hesitance to see or seek value in invested or interested ignorance is understandable, given the role that this kind of ignorance plays in oppressive epistemic practices. But perhaps it is a mistake to denigrate invested or interested ignorance entirely.

I suggest that invested or interested ignorance may stimulate and sustain practices of resistance against unjust distributions of epistemic responsibilities. I introduce the concept of *epistemic burdens* to draw attention to injustices within the social distribution of epistemic responsibilities. Epistemic burdens are not identical with epistemic responsibilities, though the two concepts are interconnected. To bear an epistemic burden is not always to bear an epistemic responsibility (in the obligatory sense of responsibility), for one may bear a burden for which one is not strictly and necessarily responsible, for example, because one freely takes on the burden through a nonobligatory act of epistemic generosity, or because one is unjustly forced to bear a nonobligatory burden against one's will. Whether an epistemic burden is experienced as a burden of *knowing*, or as a burden of *having to know*, the imposition of unwelcome epistemic burdens may inflict unjust epistemic harm upon a person "*in her capacity as a knower*" (Fricker 2007, 20). It is within the shadow of unjust distributions of epistemic burdens that ignorance's value as a source of oppositional, liberatory power against unjust distributions of epistemic responsibilities becomes visible.

In this chapter I will illustrate unjust distributions of epistemic burdens through some examples that engage feminist studies and race studies. These examples cannot exhaust the wide range of injustices involving the distribution of epistemic burdens, but they are sufficient to call attention to the value of ignorance in relation to these epistemic injustices. The discussion of epistemic burdens and the value of ignorance remains programmatic here, for there are a number of highly complex and interwoven questions that must be addressed in order to map various features of the layered social, political, and epistemological landscapes (including human and nonhuman social actors) that enable unjust distributions of epistemic burdens, but which also provide resources for challenging those injustices.

THE BURDEN OF KNOWING

Advances in science and technology produce new knowledge, and responsibilities for this new knowledge must be distributed socially, across a variety of stakeholders who have differing interests. These new responsibilities may have profound social effects. Opportunities for social advancement may arise for those who claim jurisdictional responsibility for new knowledge. Other actors may face the prospect of a loss of social status on account of a change in the epistemic landscape. Consider, for example, the field of genetic medicine. Genetic testing is becoming increasingly common, and as knowledge of the human genome increases, so too will the variety of genetic tests that can be performed. In *To Test or Not to Test*, Doris Zallen (2008) argues that genetic testing differs from other kinds of medical tests because genetic tests reveal information that pertains not only to the test subject, but also to the subject's genetic kin. For this reason, knowledge obtained through genetic testing entails epistemic responsibilities in relation to oneself and in relation to others. "The era of genetic medicine has implications for each of us," Zallen writes, "with its emphasis on prevention, genetic medicine requires us to become proactive guardians of our own health" (148). Moreover, "sharing your genetic information with your own family . . . translates to a responsibility to inform other people in the family of a positive genetic-test result, so that they will be aware that they may be at a higher risk for a particular illness" (54). Knowing about one's genetic makeup entails certain epistemic responsibilities. But do people have a responsibility to know as much about their genetic makeup as they can know?

Consider the case of one woman's experience after being diagnosed with a high genetic risk of developing breast cancer. Recalling her yearlong deliberation about whether or not to undergo a prophylactic mastectomy to reduce her risk of cancer, she states in an interview conducted by Zallen:

> I would say, okay, knowledge is power, and I started talking to a lot of people and to as many doctors as I could to get the information, but ultimately I found out it was my own decision. . . . I could not understand why it was such a difficult decision for me. . . . I felt a lot of anger, and the anger was that I had to make a decision. That is what I felt angry about, that I was faced with a decision. . . . But I couldn't understand why I was so angry with the whole thing because people always told me that knowledge is power. (79–80)

The problem recounted by this woman involves her questioning the habitual devaluation of ignorance that she perceives to be socially imperative. The refrain "knowledge is power" forces itself upon her, but a powerful emotion (anger) strains against it. Through her anger she perceives the burdensomeness of knowing, and she begins to consider the value of ignorance. Zallen observes that some people, whom Townley (2011) calls epistemophiliacs, are "information seekers for whom knowledge is power. Information itself is of value and they wish to have as much of it as possible. But for others," Zallen (2008) continues, "medical information may be far less desirable, especially if it brings with it the prospect of endless worry" (94). Recalling her long deliberations, the woman considering mastectomy focuses on the epistemic burden that knowing her genetic makeup imposes on her. Questioning the vague, social pressure to know, which initially led her to acquire knowledge of her genetic makeup, she begins to doubt whether it is always better to know. She develops an interest in being free of the *burden of knowing*. In contemporary medical contexts, increased social pressure to know more about one's health is partly a product of the intensifying moralization of medical health, whereby matters of medical health increasingly act as overriding reasons in personal and social deliberations. To question one's obligation *to know* as much as possible about one's own health is to resist the yoke of medical epistemophilia. Anger about bearing a burden of knowing may awaken an interest in resisting blind acquiescence to a vague but potent social pressure *to know*.

The Burden to Know

The principle of personal epistemic responsibility is evident not only among health-care consumers, but also in consumer culture at large, wherein individuals are expected to exercise vigilant circumspection with regard to the endless flux of products and services that come and go from the marketplace. In contemporary consumer contexts, the neoliberal mantra of personal responsibility inverts classical liberalism's astute recognition of complex social-epistemic interdependence. The epistemic burdens heaped upon individual consumers belie respect for the classical liberal giant Friedrich Hayek's fundamental epistemic principle: "Individual reason is very limited and imperfect," and it is folly to believe that "everything which man [*sic*] achieves is the direct result of, and therefore subject to, the control of individual reason" (1984, 136). Over the last forty years, philosophers, social psychologists, economists, and cognitive scientists have repeatedly buttressed Hayek's epistemic principle on empirical grounds, arguing that human beings are shockingly bad reasonsers.[4] These findings ought to make us question whether anyone is equipped to bear the epistemic burdens that are tacitly but forcibly delegated to consumers by way of what Mills and Sullivan might call a substantively ignorant market logic—a logic that deceitfully insists upon its own neutrality.

It would be unjust to require people to bear epistemic burdens that they cannot bear. But injustices in the social distribution of epistemic burdens need not have anything to do with ability directly. To illustrate the questionable distribution of epistemic burdens in contemporary consumer culture, let us turn to another example involving health-care consumers. In a *Huffington Post* article on June 21, 2011, about doctors' and clinics' frequent performance of unnecessary cervical cancer screenings, Lauran Neergard reports that a recent U.S. Center for Disease Control and Prevention (CDC) study recommends, "women have to be savvy to ensure they're getting the right checkups—enough, but not too much." Here, women are asked to be responsible, proactive health-care consumers, to assume the epistemic burden *to know* whether a cervical cancer screening ordered by a doctor is medically warranted. At first blush it may appear wise for women (or, for that matter, any group of health-care consumers) to check up on their doctors' recommendations. After all, assuming some measure of personal responsibility for the quality of one's own health care could provide an additional safeguard and empower one to protect oneself against medical paternalism. Knowledge is power. However, health-care consumers might prefer not to bear the epistemic burden of needing to check up on their doctors (see Lupton 2012), and there is good reason to respect health-care consumers' insistence that their preference *not* to bear epistemic burdens ought to be respected.

Consider the case of alerts and recommendations that are frequently issued to communities following reports of sexual assaults within those communities. All too often, women are asked/told to be aware of their surroundings, or not to walk alone, in order to protect themselves against sexual assault and rape. In these cases, women are asked to bear certain epistemic burdens—for example, to be especially alert and on guard—because they are at greater risk of being sexually assaulted. The epistemic costs of increased risks are passed along to members of a group (women) who are most at risk. The community alerts and recommendations here described blame the victim by signaling that victims of sexual assault may have been able to do more (epistemically or otherwise) to protect themselves against violent crime. One serious problem with placing the epistemic burdens of increased risk onto those who are most at risk is that this distribution fails to address the broader, oppressive social conditions that place women at increased risk in the first place. Instead of seeking to rectify the deep social injustices of patriarchy, the familiar community alert represents an additional act of *epistemic injustice*, one that encumbers women with epistemic burdens that are a result of broad patterns of patriarchal injustice.

Informing women that they ought to be savvy about their medical checkups follows the same pattern as alerting women to be more aware of their surroundings in order to avoid sexual assault. Placing upon women the epistemic burden of having to check up on their doctors fails to address the long history of medical patriarchy's oppression of women and violence against women's bodies. Instead of seeking to remedy broader social injustices that contribute to medicine's neglect and subjugation of the female body, an additional act of *epistemic injustice* is committed by encumbering women with additional epistemic burdens that are a consequence of medical injustices.

Distributing Burdens

Concern about the injustice of distributions of epistemic burdens extends far beyond the sociopolitical contexts of consumerism or sexism. Indeed, a single scene from the popular Showtime series, *Homeland*, illustrates a diversity of social relations that may sustain unjust distributions of epistemic burdens.[5]

Mortally wounded and fleeing from the U.S. government, Nicolas Brody, a white American soldier turned terrorist, is taken to an impoverished neighborhood in Caracas, Venezuela, where he is saved by a group of criminal outcasts who are likewise on the lam. Forbidden to leave the dilapidated high rise in which his new companions (or captors) dwell, Brody is constantly attended by a young Hispanic woman, Esme, who nurses Brody back to health under the direction of her domineering father, who is one of Brody's captors. In the scene of interest, Esme accompanies Brody as he takes his first tentative steps. "Ah, it feels good to be walking," says Brody. "Walking?" asks Esme who does not speak English. "Walking," replies Brody, who does not speak Spanish. Picking up a piece of PVC piping in one hand, Brody wiggles two fingers on his other hand to resemble legs, explaining, "We're walking." "We're walking," Esme responds. "Yes, we're walking. We're walking. With a cane," says Brody, tapping the piping on the ground. "Baston," Esme points out. "Baston?" Briefly distracted, Esme does not reply to Brody's question. "We're walking with a cane," Brody continues. "We're walking with a cane," says Esme dutifully. Over a short time, Esme learns to speak English with Brody, and we never see the two conversing in Spanish.[6]

The show suggests that Esme may be in love with Brody. But even if she is, her readiness to learn English is not an unproblematic act of epistemic generosity. Epistemic generosity may take the form of freely and gladly giving epistemic goods to another, as Roberts and Wood (2007, 286–304) point out, or it may take the form of freely and gladly relieving another of an intellectual burden. In either case, a purely loving generosity implies equal freedom to give or relieve or *not* to give or relieve. But the relationship between Esme and Brody is not simply that of a lover and her oblivious beloved. It is also a relationship between a young Hispanic woman, tethered to her father and living in the Global South, and an older, cosmopolitan, white man from America. These features of Esme and Brody's relational, group identities carry the legacy of histories of oppression and ongoing disparities of power that must be taken into account when evaluating the justice or injustice of the distribution of epistemic burdens in this case. Esme's readiness to speak English may not be free and glad, even if she is in love with Brody. And Brody's assumption that English will become his and Esme's common language may be a feature of a substantive epistemic practice—even if, as Sullivan (2006) points out, those practices are unconsciously habitual.

Prospects for Theory and Practice

Epistemic injustice harms people not only in their capacities as knowers, but also in their capacities as nonknowers. In the present study, I have sought to illustrate and explain that ignorance may have value for individuals and groups who have been habitually and unjustly encumbered with epistemic burdens that it is not necessarily their responsibility to bear. I have identified two types of epistemic burdens, *burdens of knowing* and *burdens to know*. As grounds for resisting epistemic burdens I have cited philosophical and social-scientific arguments for the woefully limited power of human cognition. More importantly, I have evoked histories and contemporary practices of oppression and violence—specifically against white women and Hispanic women—as grounds for challenging status quo distributions of epistemic burdens.

Further study of epistemic burdens may ask us to think about who or what may bear epistemic burdens, for the bearers of burdens could include businesses, industries, governments, institutions, communities, African Americans, Indians, consumers, women, Muslims, inhabitants of the Global South, and so on. Further study should also lay bare the various actors and mechanisms by which specific distributions of epistemic burdens are initialized, maintained, and challenged, a project that would benefit most from engagement with both historical and contemporary sources. We could strengthen the study of epistemic burdens by adopting a normative posture, asking when it is appropriate (and when inappropriate) to refuse epistemic burdens, and asking how certain epistemic burdens *ought* to be distributed. This work will be worth doing, as it can make substantial contributions to the study of epistemic responsibility, which heretofore has tended to focus on responsibility as a condition of knowledge (Code 1987; Zagzebski 1996), or as an otherwise valuable epistemic trait (Kvanvig 1992), but less on questions directly related to the social distribution of epistemic responsibilities. This work may also provide new tools for understanding the processes by which expertise is fashioned, managed, and disputed within complex epistemic communities, for example, by viewing the construction of expertise as involving the discipline of nonexperts by managing what nonexperts ought *to know*.

Perhaps ignorance could be deployed practically, intentionally, and interestedly, to expose and challenge habitual and unjust distributions of epistemic burdens, and thereby serve the ends of liberatory social epistemology. We might use the term *oppositional ignorance* to refer to intentional and interested ignorance when it is used not to distort reality, but to reveal realities of epistemic oppression. To illustrate what oppositional ignorance could look like, let us return to the case of Esme and Brody. Consider also a hypothetical case in which Esme does not know English, and maintains a commitment to her own ignorance of English—perhaps because she values, for its own sake, a life in which she is free of the burden to know English. If Esme interestedly and intentionally refuses to learn English in order to demonstrate to Brody the burdensomeness of being required to learn another's language, or simply to compel Brody to learn to speak Spanish if he wants to converse with her and seek her help, Esme is using her ignorance oppositionally. Our hypothetical Esme challenges Brody's position of privilege as a white man from the Global North, exposing Brody's powerfully ignorant presumption that it is Esme's responsibility to learn English. I hope that further discussion of unjust distributions of epistemic burdens can facilitate practical, liberatory action, perhaps in the form of collective acts of oppositional ignorance.

NOTES

1. See, for example, Philip Olson and Laura Gillman, "Combating Racialized and Gendered Ignorance: Theorizing a Transactional Pedagogy of Friendship," *Feminist Formations* 25, no. 1 (2013): 59–83; Sullivan and Tuana, *Race and Epistemologies of Ignorance*; Shannon Sullivan, *Revealing Whiteness: The Unconscious Habits of Racial Privilege* (Bloomington: Indiana University Press, 2006); Sandra Harding, "Two Influential Theories of Ignorance and Philosophy's Interests in Ignoring Them," *Hypatia* 21, no. 3 (2006): 20–36; and Mariana Ortega, "Being Lovingly, Knowingly Ignorant: White Feminism and Women of Color," *Hypatia* 21, no. 3 (2006): 56–74.

2. See also Cynthia Townley, *A Defense of Ignorance: Its Value for Knowers and Roles in Feminist and Social Epistemologies* (Lanham, MD: Lexington Books, 2011).

3. See Philip Olson, "The Virtues of Flourishing Rational Agents," PhD diss., Emory University (2006); and Philip Olson, "Putting Knowledge in its Place," *Philosophical Studies* 159, no. 2 (2012): 241–61.

4. See, for example, Amos Tversky and Daniel Kahneman, "Judgment under Uncertainty: Heuristics and Biases," *Science* 185 (1974): 1124–31; Christopher Cherniak, *Minimal Rationality* (Cambridge: MIT Press, 1986); and John Doris, *Lack of Character: Personality and Moral Behavior* (Cambridge: Cambridge University Press, 2002).

5. *Homeland*, Season 3, episode 3, directed by Clark Johnson. New York; Showtime, 2013. Television series.

6. It would be incorrect to think it natural for a U.S. television show to slip into English in this scene, especially since *Homeland* regularly subtitles scenes in which languages other than English are being spoken.

BIBLIOGRAPHY

Alcoff, Linda M. "Epistemologies of Ignorance: Three Types." In *Race and Epistemologies of Ignorance*, edited by Shannon Sullivan and Nancy Tuana, 39–58. Albany: State University of New York Press, 2007.

Axtell, Guy. "Expanding Epistemology: A Responsibilist Approach." *Philosophical Papers* 37, no. 1 (2008): 51–87.

Cherniak, Christopher. *Minimal Rationality*. Cambridge: MIT Press, 1986.

Code, Lorraine. *Epistemic Responsibility*. Hanover: University Press of New England and Brown University Press, 1987.

Doris, John. *Lack of Character*. Cambridge: Cambridge University Press, 2002.

Fricker, Miranda. *Epistemic Injustice*. Oxford: Oxford University Press, 2007.

Harding, Sandra. "Two Influential Theories of Ignorance and Philosophy's Interests in Ignoring Them." *Hypatia* 21, no. 3 (2006): 20–36.

Hayek, Friedrich. "Individualism: True and False." In *The Essence of Hayek*, edited by Chiaki Nishiyama and Kurt R. Leube, 131–59. Stanford: Hoover Institution Press, 1984 [1948].

Homeland. Directed by Clark Johnson. Written by Howard Gordon, Alex Gansa, Gideon Raff, Henry Bromell, William Bromell. Performed by Damian Lewis (as Brody) and Martina Garcia (as Esme). Showtime, October 13, 2013.

Kvanvig, Jonathan. *The Intellectual Virtues and the Life of the Mind*. Savage, MD: Rowman and Littlefield, 1992.

Lupton, Deborah. *Medicine as Culture*. Thousand Oaks, CA: SAGE Publications, 2012.

Olson, Philip. "The Virtues of Flourishing Rational Agents." Unpublished doctoral dissertation, Emory University, 2006.

Olson, Philip. "Putting Knowledge in its Place." *Philosophical Studies* 159, no. 2 (2012): 241–61.

Olson, Philip and Laura Gillman. "Combating Racialized and Gendered Ignorance: Theorizing a Transactional Pedagogy of Friendship." *Feminist Formations* 25, no. 1 (2013): 59–83.

Ortega, Mariana. "Being Lovingly, Knowingly Ignorant: White Feminism and Women of Color." *Hypatia* 21, no. 3 (2006): 56–74.

Roberts, Robert C. and W. Jay Wood. *Intellectual Virtues: An Essay in Regulative Epistemology*. Oxford: Oxford University Press, 2007.

Sullivan, Shannon. *Revealing Whiteness: The Unconscious Habits of Racial Privilege*. Bloomington: Indiana University Press, 2006.

Sullivan, Shannon and Nancy Tuana, eds. *Race and Epistemologies of Ignorance*. Albany: State University of New York Press, 2007.

Townley, Cynthia. "Toward a Revaluation of Ignorance." *Hypatia: A Journal of Feminist Philosophy* 21, no. 3 (2006): 35–77.

Townley, Cynthia. *A Defense of Ignorance*. Lanham, MD: Lexington Books, 2011.

Tuana, Nancy and Shannon Sullivan, eds. "Special Issue: Feminist Epistemologies of Ignorance." *Hypatia: A Journal of Feminist Philosophy* 21, no. 3 (2006): vii–ix.

Tversky, Amos and Daniel Kahneman. "Judgment under Uncertainty: Heuristics and Biases." *Science* 185 (1974): 1124–31.

Zagzebski, Linda. *Virtues of the Mind*. Cambridge: Cambridge University Press, 1996.

Zallen, Doris. *To Test or Not to Test*. Piscataway, NJ: Rutgers University Press, 2008.

Chapter 12

Freeing Knowledge

The Future of Critical Knowledge Production in the New Age of Corporate Universities and the Renegade Generation of Researchers

Adam Riggio

Of all the original Collective Vision statements first published on the Social Epistemology Review and Reply Collective (SERRC) website, mine was one of the earliest, and perhaps one of the worst, a hodgepodge of notes about my current projects. Since then, much has changed in my own situation as a scholar. Despite above-average performance in research and publication, regular participation in an international scholarly community through conferences and the collective, and regular teaching work throughout my doctorate, three years on the university job market has seen me unable to secure even a paltry per-course position, not even for a single term. As this book went into its planning stages, I was deciding to forgo a university career for a new path in Canada's not-for-profit sector.

My story has become sadly typical of many in my generation of scholars, and as I write this chapter, I know that it will also be the fate of many SERRC members who are still working through their graduate studies. Even researchers at the most prestigious universities are losing their research capacities and even their jobs as government funding for pure research dries up and the postsecondary sector relies increasingly on private industry partnerships that are hostile to critical work or research that cannot be immediately monetized. As well, these corporate partners change the dynamic of university instruction from critical education to a customer service experience that privileges student satisfaction instead of challenging accepted beliefs.

THE END OF THE INSTITUTIONAL ERA

The current labor crisis in the contemporary university system is quantitatively worse than ever. An era of austerity in public services, including universities, has combined with an academic culture that encourages and produces PhD students with the needs of departments for teaching assistants and other cheap academic labor at the forefront. The result has been an overproduction of people qualified to teach and research in universities, often who are trained to work only in this capacity, at the same time as the number of dignified positions such people can fill has plummeted. The result is chronic under- and unemployment that drives bright, creative people out of humanities research. I call this the Lost Generation of the university system. If university researchers are the paramount producers of innovative, critical knowledge in our society, then the continuing adoption of corporate governance and management norms in the higher education sector will prevent the creation of innovative, critical knowledge. The public research institutions where the critical disciplines of the humanities and social sciences have settled no longer give such knowledge a welcome home.

The question for my generation is whether we, as producers of critical knowledge, should give up as we move on to other fields of work that can still offer us dignified lives. If the university as an institution is no longer interested in knowledge that addresses social injustices and challenges the status quo, then it is perfectly sensible to treat this as another failed career path. The humanities will be an intellectual rust belt and my generation will be its casualties, abandoning the university just as the university abandoned us. This would be a sensible path, but it would be mean in spirit, resentful, and petty. The best of the Lost Generation shares a passion to create knowledge that is uncomfortable for the powerful and develops the concepts that drive the moral progress of our societies. For decades, the university had been the institution that housed and encouraged this progressive process. This is no longer the case, or at least it will likely not be for much longer.

But one does not abandon a noble cause because the easiest means of working toward it disappears. If one is strong enough, then the means must change while the end remains.

THE LIMITS OF REFORM IN A VULNERABLE SYSTEM

It should be clear at this point that my current Collective Vision contribution has quite a different style than most of the others. It takes the form of a manifesto, or to be less pretentious, at least a polemic or a well-informed opinion. The opinion piece and the polemic are standard forms of writing in the arsenal of the public intellectual, the gadfly of critical knowledge production. If we believe Henry Giroux, for example, a university-based public intellectual of an older generation, that a university post is necessary for such a public critic and conscience, then a neoliberal model will destroy this role.[1]

For all Giroux's righteous rage, the position he advocates is ultimately a conservative one. He argues that the university must be preserved as the ideal place for progressive voices and the associated knowledge production. Ultimately, it is a defensive argument, and as such inherently disadvantaged: he argues for the retrenchment of a position that we know to be vulnerable. The role of the university as the safe home for critical public intellectuals is itself under threat because of a massive cultural and economic movement that has already destabilized the safety of this institution for progressive knowledge production. I think the battle to save the university as an institution from the takeover of corporate governance norms has already been lost. Let's not mourn what is gone for too long. Even if I thought the university system could be repaired as Giroux dreams, it would remain vulnerable to the same assaults as it has faced over the last several years: governments can continue to cut public university funding and corporate partners can prioritize equity investments over education and research, while tuition dollars would be repurposed as liquidity instead of direct funding.

These vulnerabilities to corporate assault aside, the university is not even the truly best place to produce progressive public knowledge. The current organization of our universities isolates knowledge producers and products of humanities disciplines from the public that they should ostensibly engage if the vision of researchers as public intellectuals holds. Professionalization norms, styles of discourse, the nature of the audience for original research in those styles, and the physical accessibility and prominence of our publication venues all contribute to the isolation of progressive knowledge from public discussion.

The norms of professionalization in research universities are the standards for promotion and tenure. The research works that carry the most weight in these evaluations are publications in peer-reviewed disciplinary journals. Peer review norms and methods are continually subject to critique and ideas for refinement and change. Some of these have come in public venues from famous figures in social epistemology, such as Steve Fuller, who in 1999 hosted an entire conference on the subject. But the most important problem lies in the audience of such journals, other professional practitioners of the same discipline or subdiscipline. All the readers share the technical vocabulary, background assumptions, and historical knowledge of the discipline, or else they are students learning these exact things to become professional practitioners. Their professional status rests on being integrated into a research university institution. Whether one is integrated into such an institution depends on one's success in publishing in the most prestigious disciplinary journals.

Such success all too often requires tailoring the problems one works on and the terms one uses to discuss them to the technical vocabulary and style of journals whose only audience is other experts. While there are exceptions to this exclusionary publication pressure (e.g., I know one PhD student at University of Guelph in Canada whose artistic exhibitions count as professional publications because her dissertation topic discusses philosophy's relations to the visual arts), they remain exceptions. All too often, the academic publications recognized to have the highest professional value require a style of identifying, framing, and discussing problems that leave a wider public alienated.

Knowledge developed in this heavily professionalized context may be ostensibly open through public libraries being able to access otherwise restricted university holdings, but the public is not an engaged partner in disciplinary knowledge creation. Publications in disciplinary journals tend to shrink the range of their focus to increasingly narrow debates and controversies, arguing over technical concerns whose relation to the daily lives of most people is usually tangential at best. The only people who would know what is even going on in a given article in the top disciplinary journals of a university-based research profession are other professional researchers. The general public would be entirely alienated from this discourse.

There have been attempts over the last decades to improve the physical accessibility of these publications as well, to expose the public to these journals, even as their style would put more people off than engage them in the issues they cover. A multidisciplinary boycott of prominent journal publisher Elsevier has been ongoing since 2012 because of that company's prohibitive subscription prices.[2] Open access journals are often touted as an alternative to the high costs of disciplinary journal subscriptions, but they face two intractable problems. The first is that open access journals are not regarded with suitable prestige in tenure and promotion processes to encourage university-based researchers to focus their writing efforts there. The second is that open access journals are too often funded by soliciting the authors themselves to pay for their own publication. Such pay-to-publish exchanges are not possible for many academic professionals, especially younger ones who are frequently trapped in poorly paid adjunct labor positions. The pay-to-publish model is also detrimental to developing the prestige of a journal because it is essentially the same financial framework as vanity press publishing.

Another casualty of the increasingly corporatized model of university management has been the slashing of library budgets for acquisition and journal subscriptions. Although public libraries often partner with local universities for free public access to disciplinary journals, such a practice would be increasingly difficult for a community library in a city without a university. Given the tough choices that plummeting budgets force upon many research institutions regarding their own library holdings, such a strategy for public accessibility would have a geographically limited scope of success. And it would still not approach the biggest problem that heavily disciplined professional journals face: they would still be written for professionals only, and not the public that humanities research ostensibly serves.

EXPERIMENTAL COMPOSITION IN THE HUMANITIES

Some members of the Lost Generation do gain reasonable employment in the university system after a while, despite the pressures of a corporatized university institution that prefers lowering teaching labor costs by expanding adjunct positions and experiments in exploding class sizes through tools like Massive Open Online Courses. Thankfully, the most idealistic members of the Lost Generation are developing new ways to create knowledge without the institutional support of universities. These new knowledge products have more potential to connect with a public in our troubled times than the strictly disciplined knowledge products of professional journals and the highly technical books that flow from journal-dominated modes of knowledge production. I will describe three techniques that I am developing in my own career as a burgeoning public intellectual whose professional home lies outside the university system: popular nonfiction, blogs, and the arts.

One avenue for future progressive works strongly resembles many of the works produced today: popular nonfiction. Empirical, historical, and literary research can unite with the conceptual power of philosophical reasoning, analysis, and rhetoric to craft works that deal with the major problems of our time in frank popular discourse. The university researcher can leave an anemic job market for a career as a popular author seeking to shape public consciousness and political action around the major social problems of our time. For example, I am currently shopping a manuscript developed from my dissertation to nonfiction publishers, a philosophical analysis that combines the moral concepts in environmentalist moralities and political activism with the scientific principles underlying contemporary ecological and biological sciences. This is one way to put some conceptual depth to the rhetoric of interconnectedness and interdependence that dominates environmentalist discourse. Because our selves and our world are constituted from the constant interaction of increasingly complex feedback loops of affects, an environmentalist morality is the only framework to build a genuinely sustainable technology.

A book on subjects such as ecology, climate change, the internal and external challenges democracy faces, global politics, and the histories of nations and peoples can strike a chord with a wide audience. Combined with a strong knowledge of public relations and media promotion techniques, an author of nonfiction can build an audience for works that are written in a tone and style acceptable for public consumption. These books would not, perhaps, immediately gain the recognition of university-based academics. I have often encountered pretensions to superiority over the public among friends, professionals, and colleagues who are still embedded in the university system. Popular nonfiction would not stand up to the technical rigor of proper disciplinary publication, and I can understand the perspective of such a hypothetical member of the older generation of university researcher who would consider popular publication beneath him. Yet, it is the popular books that actually have an impact on public opinion, discourse, and action, actually contributing to the political movements that constitute social progress.

The second model for progressive works in humanities research emerging from the Lost Generation is the most radical. This new form of publication depends on the Internet as its home. I call it the epic blog, a research, and interpretive project so enormous and multifaceted that it can only unfold in short chapters of a couple of thousand or so words at a time, each individual facet spread out over a period that can take years to unfold completely, even with two or three publications per week. The result is a single work of scholarship approaching Proustian dimensions, or even greater. Two central examples have arisen in the discipline of literary/media studies, in the works of Philip Sandifer[3] and Joshua Marsfelder.[4]

Sandifer is the creator of TARDIS Eruditorum, a four-year project examining individually every *Doctor Who* story transmitted on television, as well as ancillary material from other media and essays on cultural developments and relevant concurrent artistic products and movements. He has also begun an even more sprawling project, *The Last War in Albion*, on the influx of British creators into American comics over the 1980s and their influence since then, with a particular focus on the relationship of Alan Moore and Grant Morrison. These projects all generate enough income for Sandifer to live, through sales of individual books adapting and expanding the online essays, as well as Kickstarter and Patreon backing from his readers directly. Since 2013, Marsfelder has run a similar project on the Star Trek franchise, incorporating as well an impressively idiosyncratic analysis of indigenous forms of knowledge (particularly the mythic and scientific cultures of ancient Polynesian oceanic navigators). My own blog currently exists more as a promotional exercise, a similarly sprawling project of analysis of my ongoing research and artistic work. It is a kind of paratext or living commentary track on the research and writing that informs my in-progress works for publication.

Blogs are also increasingly important in activism to change the university institution itself. One example lies in another significant story of the Lost Generation of university researchers: the marginalization of female knowledge producers, even compared to other males within the same disadvantaged demographic. My own discipline of philosophy is an especially egregious offender, notable examples of hostility, harassment, and violence toward female practitioners being Colin McGinn, Peter Ludlow, and the culture of the entire philosophy department at the University of Colorado's Boulder campus. I do not tell this story here because I do not consider myself entitled to speak for the many women who have suffered harassment or the thousands of small indignities of trying to find a place in a professorial occupation whose employment norms remain defined according to an era before two-career families became normal. The blogs "Women in Philosophy"[5] and "What We're Doing About What It's Like"[6] are far better chronicles than I can be in this essay of my discipline's gender problems in its university sector.

Knowledge production is inherently a creative activity, and the humanities themselves have long had a special relationship and a parallel social role with the arts. The most legendary and respected artistic products interpret human life and politics, and point the way toward progressive social change. Nonfiction publications are themselves inherently creative, and their conception, development, and composition require an artistic sensibility. Humanities education has become so widespread, and its basic interpretive ideas so prevalent in our culture, as evidenced in the growing popularity of think piece journalism, that many casual viewers have similar interpretive abilities of art and media as a professor of literary and media studies. So it is only natural that the Lost Generation turn their philosophical and interpretive sophistication to the creation of art that can engage with the complex aesthetic powers of the modern audience.

My own work is the best example I can think of at the moment. While our essays for this volume were being assembled, I had my debut as a playwright in an independent theater in my current home of Hamilton, Ontario. The characters of *You Were My Friend* are studies in Nietzsche's conceptions of virtue: one develops a personal strength and quiet nobility to overcome the adversity she faces, while the other is a case study in his image of the weak, resentful soul. My debut novel *Under the Trees, Eaten,*[7] published at the same time, evokes, critiques, and parodies the tropes and underlying ideas of the work of H. P. Lovecraft. Its alien creatures are named The Diggers, a name that evokes English communist societies in the Cromwell era, and explores what kind of morality would develop among creatures with no sense of individuality. None of these projects would be possible in their current forms without my professional-level knowledge of philosophy.

RELISH OUR NEW DIVERSITY: THE RENEGADE GENERATION

At our best, humanity handles crisis through creativity and regeneration. The university as an institution once offered a secure home where research specialists could produce and disseminate knowledge to progress and improve the human situation materially, technologically, maybe even morally and spiritually. The corporate business model's corruption of that ideal of public service has perhaps irreparably harmed the university's capacity for social progress. It is all too easy to think of this demographic of surplus professional university research and teaching labor as a Lost Generation, whose lack of institutional support will see their talent wasted.

I will not respond to this economic and demographic crisis with resentment and resignation. If the newly corporatized university institution betrayed the Lost Generation of researchers who believed its promises of a dignified future, then I accept its challenge to become a greater creator of knowledge and public intellectual than that institution could have enabled. Instead of a Lost Generation, we can be a Renegade Generation, creating new flexible and decentralized institutions to create and disseminate critical knowledge that will not be vulnerable to the temptations of corruption and compromise with neoliberal values.

Conscience and dedication to freedom and progress must be more powerful than institutionalization, empty discipline, and hierarchy.

NOTES

1. Henry Giroux, *Neo-Liberalism's War on Higher Education* (Toronto: Between the Lines, 2014).
2. See the continuing petition at thecostofknowledge.com.
3. philipsandifer.com.
4. vakarangi.blogspot.com.
5. beingawomaninphilosophy.wordpress.com.
6. whatweredoingaboutwhatitslike.wordpress.com.
7. Adam Riggio, *Under the Trees, Eaten* (London, Ontario: BlankSpace Publications, 2014).

Part III

REGARDING THE INDIVIDUAL AND THE COLLECTIVE

Chapter 13

Are You Thinking What We're Thinking? Collective Visions, Group Knowledge, and Extended Minds

Eric Kerr

Most accounts of what social epistemology is and what social epistemologists do will include, in some form, an analysis of how knowledge is produced within a group or community: a collective, in other words. This could mean, in its weakest form, the ways in which individuals depend on the efforts of others for their own knowledge or, in its strongest form, the ways in which groups, rather than individuals, become the bearers of knowledge. The latter meaning has received a lot of attention recently among epistemologists and philosophers of mind (Giere 2002a, b, 2007; Gilbert 2004; Goldberg 2010; Goldman 2004; Hakli 2007; Lackey 2012; Lahroodi 2007; List 2005; Mathiesen 2006; Palermos and Prichard 2013; Pettit 2003; Tollefsen 2002, 2007; Tumollini and Castelfranchi 2006; Tuomela 2004). We often talk about groups having both knowledge and visions. In this volume specifically, we speak about a collective vision. Taking such talk seriously, can this research lend support to a strong view of collective visions?

SOME ORDINARY OBSERVATIONS

The idea has received renewed attention in recent years. Sometimes this is motivated by concerns that our common categories of what kind of organism (or organization) can know things—viz. individual biological organisms—is unduly limited (Gilbert 1989, 2004; Hutchins 1995, Pettit 2003; Rupert 2005). Call this the metaphysical problem of group knowledge. Other times it is motivated by observations about ordinary language use (Gilbert 2002; Goldman 2004, 12; Hakli 2007, 249; Quinton 1975/1976, 17; Schmitt 1994, 257–58)—viz. that we often talk as if groups know things. Call this the ordinary language problem of group knowledge. Let's begin with the latter by considering the following headlines:

BP and Halliburton knew of Gulf oil well cement flaws.[1]

Google knows you better than you know yourself.[2]

We are familiar with this kind of talk but, on reflection, it sounds a little odd. BP, Halliburton, and Google are corporations and therefore as incapable of knowing as they are of hoping. We do call corporations *persons* but this is a legal artifice. Corporations do not, and cannot, *know* anything, just as they cannot hope, wish, think, dream, believe, or imagine anything. Perhaps this is because knowing requires having a mind that requires having a brain and a body. Across disciplines, and particularly since the rise of biology in the eighteenth and nineteenth centuries, people have used the metaphors of brains, cognition, behavior, and organism to talk about groups, organizations, cities, societies, and so on. For the most part, such claims stay in the realm of metaphor and the claim that a city, for example, could *know* something sounds like a metaphor stretched to breaking point.[3] So who or what are we referring to when we say a collective knows? It seems like there are a few possible answers:

1. We don't mean that the collective (legal entity, the corporation, the group) knows. It is a metaphorical way of talking about collectives *as if* they were organisms. It can be useful to do this when you want to talk about the actions taken by a corporation or its effects on the world. Call this the Metaphorical View.
2. We don't mean that the collective knows. It's just a useful shorthand for saying that an appointed representative or authority within the corporation knows. Companies have CEOs, directors, and so on who make decisions on behalf of the whole organization and talk about "Google's vision" is a way of talking about Larry Page or Eric Schmidt's (or someone else's) vision. For example, a team of scientists may have worked on a project that resulted in some new knowledge. Not every individual would know the results of the project but there would be someone who does and when we say the collective knows we are attributing knowledge to this individual. Call this the Spokesperson View.
3. We don't mean that the collective knows. We mean that the individual members of the collective know. This could be all the members or just one individual if they are "weighty" enough; it could be a majority or it could be some other quorum. Call this the Aggregation View.
4. We do mean that the collective knows. Knowledge does require mental states such as beliefs but some groups have mental states under certain conditions. Call this the Extended Mind View.
5. We do mean that the collective knows. Knowledge does not require mental states such as beliefs and our ordinary understanding of knowledge

unduly limits knowers to only those who also have beliefs. Call this the Beliefless View.
6. We mean that the collective is the proper bearer of epistemic responsibility. Under certain conditions it is more appropriate to credit the group rather than the individual or any set of individuals with acquiring knowledge. Call this the Group Credit View.

Let's look at each of these in turn. We'll see whether they give social epistemologists who wish to make sense of either the metaphysical problem or the ordinary language problem any hope and whether it can make sense of the idea of "collective visions."

THE METAPHORICAL VIEW

At first glance, the simplest explanation is that we are speaking metaphorically. Of course groups don't know things, but talking about them as if they do can be useful for explaining, describing, or predicting certain events that the group might cause or be otherwise involved in. We often attribute mental states to groups but it is useful to note that not just any metaphor will do. No one has ever said, for example, that Google is depressed or that BP experienced great joy following the wedding of two of its employees. At least, if someone did say such a thing, it would be clear they were not talking literally. Joshua Knobe and Jesse Prinz (forthcoming) distinguish between phenomenal and nonphenomenal states. Phenomenal states for some agent, S, include "S feels depressed," "S experiences joy," "S feels happy," "S feels pain," "S feels angry," "S feels scared." Nonphenomenal states include "S intends," "S decides," "S tries," "S wants," "S believes," "S hopes," "S loves," and "S hates." We are far more likely to ascribe nonphenomenal states to groups than phenomenal ones. A Google search revealed that nonphenomenal states attributed to Microsoft numbered up to 135,000 (for "Microsoft wants") whereas they found mostly no hits at all for phenomenal states attributed to Microsoft.[4]

What about "BP knows"? Is this metaphorical? This answer would solve a lot of our worries, however, Jennifer Lackey (2012) argues that the force of this explanation may depend on the context in which the attribution is made. If, for example, I'm having dinner with a friend and tell her that on my way home a cat was walking in front of me and that "it seemed to know where I was going and was leading me to my house," my friend may well interpret me as speaking metaphorically: I don't literally think that the cat knew where I lived or that it had any intention to lead me there. If I was speaking in court, however, where, for some reason, there may be issues over whether the cat had visited my house before, then I may be pressed to clarify my statement, and I might rephrase it in less ambiguous terms or I might clarify that I really did think the cat knew where I lived. We can't reasonably say the same thing about uses such as the headlines above.

If such statements were made in court, where issues of legal and moral responsibility would be central, the reasonable interpretation could not be metaphorical. Lackey (2012, 4) notes that group knowledge attributions are made "systematically and in a widespread fashion even in contexts where there is a heightened concern for speaking precisely." There is more to it than metaphor.

THE SPOKESPERSON VIEW

Another option is to say that such attributions are a shorthand for saying that a particular individual within the collective knows the proposition in question. According to this view, when the journalist writes about Google's vision, he is really talking about Larry Page's vision (or some other member who has the authority to speak on behalf of the corporation). When we talk about what BP knew prior to the Deepwater Horizon disaster, we are really talking about what Tony Hayward, for example, knew. The Spokesperson View works well for some types of groups—chiefly those where the collective judgments are always those of some antecedently fixed group member (Pauly and van Hees 2006). However, it does not appear to work at all for groups where there is no clear individual who fits this description. When we attribute knowledge to groups we do not always refer to the kind of group where there is a clear individual who fits the description of spokesperson—for example, when we attribute knowledge to intergovernmental scientific projects.

THE AGGREGATION VIEW

Aggregation (or summative) views state that when we attribute knowledge or beliefs to groups we are attributing it to its members (Quinton 1975/1976, 17; see also Gilbert 1994; List 2005; List and Pettit 2002, 2004). In other words, we are asserting that the beliefs of some or all of its members are believed by the group as a whole. Again, such a view sounds promising but runs into several problems. Firstly, there are circumstances under which a group may not achieve consistent collective judgments even when all group members hold individually consistent judgments (Pettit 2003). List describes a "discursive dilemma" under which a majority of members of a group endorse a proposition, p, and the premise $p \rightarrow q$, but do not endorse q. The dilemma is rendered thus:

	p	$p \to q$	q
Alice	TRUE	TRUE	TRUE
Bob	TRUE	FALSE	FALSE
Carol	FALSE	TRUE	FALSE
Majority	TRUE	TRUE	FALSE

List and Pettit (2002) show that no aggregation procedure can generate consistent collective judgments (i) for any logically possible combination of complete and consistent individual judgments on the propositions, (ii) where each individual judgment has equal weight in determining collective judgments and (iii) the collective judgment on each proposition depends only on the individual judgments on that proposition, and the same pattern of dependence holds for all propositions. There are alternatives available to those who wish to preserve aggregation procedures but we won't pursue them any further here. One strong argument that they may be flawed is that it sometimes seems appropriate to attribute knowledge to groups where none of the individuals believe the proposition in question.[5]

THE EXTENDED MIND VIEW

We now move onto some of the positive answers to our original question. The first, which we might call the Extended Mind View, is that some groups have beliefs in an analogous way to individual minds. The idea has been pursued extensively (e.g., Gilbert 2004; Goldman 2004; Hutchins 1995; Pettit 2003; Rupert 2005; Theiner 2009, 2010) and I won't regurgitate the arguments for and against here. It has a certain intuitive force that can be made plain if one considers the Parity Principle due to Clark and Chalmers (1998): "If, as we confront some task, a part of the world functions as a process which, were it done in the head, we would have no hesitation in recognizing as part of the cognitive process, then that part of the world is (so we claim) part of the cognitive process" (8).

Clark and Chalmers discuss this as it pertains to technical artifacts that can extend certain cognitive processes but it is still compelling when applied to other individuals (Theiner 2009). If, in other words, a process carried out by two or more individuals is analogous in the right way to a process carried out by one individual, there is no obvious reason why one should draw a line around the individual and call that which takes place within its limits "cognitive" whereas anything outside of that line is to be called "noncognitive." Again, the arguments for and against this position are far more extensive and subtle than can be addressed here.[6]

THE GROUP CREDIT VIEW

What I am here calling the Group Credit View is beginning to be explored by epistemologists interested in the implications of the Extended Mind thesis for epistemology and, in particular, for credit theories, virtue reliabilism, and the ability intuition (e.g., Carter et al. 2014; Goldberg 2010; Hetherington 2012; Kelp 2013; Kerr and Gelfert 2014; Palermos and Pritchard 2013; Pritchard 2010, Vaesen 2011). The central idea here is that if you think that knowledge is something to do with an ability, achievement or is, in the relevant sense, creditable to an agent then it looks like you are committed to either endorsing group knowledge or distinguishing group and individual knowledge because of "biological" (i.e., nonepistemic) considerations. Given that there are many instances where it does not seem appropriate to credit any individual within a group with knowledge (e.g., her knowledge depends too much on the cognitive work of others) and yet it does seem that *someone* or perhaps *something* should be credited with knowledge, then it looks like there are also many instances of group knowledge. Again, due to space restraints I cannot go further into this complex debate but will instead point the reader in the direction of the research cited above.

DO GROUPS HAVE VISIONS?

Some account explaining both the metaphysical concerns outlined above and the ordinary language attribution of knowledge and mental states to groups is required and we are only in the early stages of an interdisciplinary research project that seeks to address both issues. While explanations for group knowledge, belief, cognition, and so on, abound, so far none of that research has considered how plausible these arguments are in relation to a specific type of mental state: visions.

Visions relate to knowledge in several ways and considering the formation of group visions may be helpful for assessing the problems of group knowledge. Visions are a kind of mental imagery sometimes referred to as "quasi-perceptual experience." In essence, a vision is similar to other perceptual experiences, but occurs in the absence of the appropriate external stimuli (Thomas 2013). Visions depend on knowledge. In order to generate a vision, one must have knowledge of the ways in which things might be. In short, one must pick out a preferred alternative from many and conjure up an image of what that alternative is. For example, I might have a vision of my career in ten years' time. In order to have this, I must first know something about my career now and I must know the possible paths my career might take over ten years.

As with group knowledge, we often attribute visions to groups. We might talk about the U.S. government's vision for society or a trade union's vision for the future of their profession. Are any of the above options of use when trying to explain what a collective vision is? Certainly we often talk metaphorically about group visions. However, as we saw with group knowledge attributions, this kind of talk can be quite systematic and made in formal contexts where linguistic precision would be required. The spokesperson analysis also looks promising. Often the vision of a group is a shorthand for the vision of a charismatic leader such as when people talked about Apple's vision as a shorthand for Steve Jobs's vision. As with group knowledge, however, this analysis makes sense only for a subset of groups. Spokesperson accounts will be difficult to apply to perceptions that are not made "vocal." It would certainly be possible for each "visionary" to sign some document endorsing a particular statement of the vision but this would be an additional "mission statement" and not the vision itself. Perhaps the Extended Mind analysis can offer something? The Extended Mind thesis (and related theories such as Embodied Cognition) have been considered in relation to perception (Clark 2008a, 169–79, 2008b; Kerr and Gelfert 2014; Ward and Stapleton 2012; Wilson 2004, 2010) but here we are concerned with "seeing in the mind's eye," "visualizing," "envisioning the future," "imagining the feel of," and so on. If anything, this would be the one kind of perception that is immune to extended mind-type arguments. After all, what marks mental imagery out as distinctive is precisely that it takes place, and can only take place, within the head and that it's specifically about things in the head not things outside the head. The charge of bioprejudice therefore loses its force.

So here we have a problem. On the one hand, while it seems fairly plausible that knowledge, belief formation, and many other kinds of mental states could be socially extended, it seems implausible that visions could. And yet it is very common to talk of collective visions. Nor is it far-fetched to suppose that one or more individuals can envision the same goal. Many activities undertaken by two or more individuals can only be pursued, in fact, under this assumption. You and I can only move the piano, for example, if we have a shared vision of where it is going to go. Of course, there are the phenomena of mass hallucinations (if such things exist) and the more verified phenomena of shared psychosis (sometimes called folie à deux) where symptoms of a delusional belief are transmitted from one individual to another. But we are looking for something more substantial than a hallucination and transference of a belief doesn't quite seem to do the job either.

Perhaps the most promising area for our purposes is psychological and philosophical research on joint intentions. This is the phenomena where one or more individuals both or all intend to perform a particular activity (such as playing chess or figure skating or moving a piano). Here, agents envision goals and set about, jointly, to realize them. According to Michael Bratman, for two individuals to have a shared intention to perform a particular activity, *J*, each of us must intend that we (continue to) *J* (Bratman 1992, 1993). Olle

Blomberg (2011) considers the idea that such intentions are subject to an "exclusivity constraint." According to this principle, one cannot intend to perform another agent's action, even if one might be able to intend *that* she performs it. For example, while I can intend *that* you leave my company at 8 p.m., I cannot intend *to* leave my company. Blomberg argues, in contrast, that there are some cases where one may intend to perform an activity that belongs both to oneself and to another agent.

He does this by first noting that intentions-in-action can be technologically extended. For example, a blind man using a cane extends his ordinary capacities through the technology of the cane. The cane becomes "transparent" in the sense that he is no longer aware that he is using it as a tool in order to find out what the ground is like (Merleau-Ponty 1945, 165–66; Maravita and Iriki 2004). Blomberg argues that a similar "phenomenon of transparency" is present in joint activity. Consider two skilled ice skaters, Alice and Bob, performing a figure dance. Here, it is not that Bob experiences himself as being in a position to determine Alice's actions. Rather, the experience is a "joint interface with the world" (Seeman 2009, 508). Just as, in the case of technologically extended activities, the tool user is no longer aware of the affordances of the tool, in socially extended activities, the agents are no longer aware of what they can intend the other does, but are jointly intended to do it.

In writing a series of "Collective Visions" it does not seem very plausible that we would ever reach the level of joint intention attained by the figure skaters. Academia thrives on dissensus, after all. However, neither is it the case that we are entirely disconnected minds and brains with our own self-contained "visions." By being part of the collective of social epistemologists we cannot fail to be involved in and influenced by the joint and individual activities going on around us even if we do not know precisely where they are headed. At the same time, we are aware from psychological studies on conformity and peer pressure that we modify our own perceptions and visions to align with a majority (exemplified in Solomon Asch's experiments in the 1950s).[7] Since this is intellectually undesirable we might be tempted to avoid collective vision. We are left with a tension between wishing to engage in a collective vision and wishing to preserve our own epistemic independence. Both are admirable goals, but we should be wary of falling too far to one side or the other.

NOTES

1. Suzanne Goldenberg and Julia Kollewe, "BP and Halliburton Knew of Gulf Oil Well Cement Flaws," *The Guardian*, 29 October 2010. www.guardian.co.uk/environment/2010/oct/29/bp-oil-spill-bp.

2. James Carmichael, "Google Knows You Better Than You Know Yourself," *The Atlantic*, 19 August 2014. www.theatlantic.com/technology/archive/2014/08/google-knows-you-better-than-you-know-yourself/378608/.

3. The idea that groups can have the kinds of doxastic capacities that individuals have (i.e., that they can know, belief, intend, and so on) has a long history which I can't do proper justice to here. Thomas Hobbes wrote about "group minds" in *Leviathan* in the seventeenth century, and the nineteenth-century physicist Gustav Fechner talked about "collective consciousness." Herbert Spencer described society as a social organism and Émile Durkheim (1893) wrote about collective consciousness in his "Division of Labour in Society" although he was clear that he did not want to reify the concept. The entymologist William Morton Wheeler (1911) was perhaps the first to draw an analogy between the behavior of individuals cooperating and a single organism. He called this cooperative interaction a "superorganism." The idea has become a common trope in science fiction literature. H. G. Wells's "world brain" and Olaf Stapledon's *Last and First Men* (1930) were early imaginations of knowledge belonging to groups rather than individuals and *Star Trek*'s Borg, perhaps the most well known. George Orwell later popularized the idea of groupthink in 1984. More recently, developments in information communication technologies (ICTs) have moved the discussion out of academia and science fiction to popular reflections on the "Wisdom of Crowds" (Surowiecki) and to understanding the Internet as a giant brain (Andrews 1986). In 2006, MIT established a Center for Collective Intelligence.

4. Here, it is worth voicing an objection to Knobe and Prinz's study. The way one couches one's search terms is critical here. If one searches for "Microsoft was happy" rather than "Microsoft feels happy," one gets around 76,200 hits (at time and place of writing), presumably because the former is a much more natural way to convey a similar meaning. Even so, one cannot do this with all of Knobe and Prinz's examples and it does seem that we are more comfortable attributing some mental states to groups and not others. "Microsoft envisions" was not part of their study, although my own search returns around 12,300 results, so it seems "envisions" might be closer to the nonphenomenal group than the phenomenal.

5. In Kerr and Gelfert (2014) we discuss a number of circumstances where groups can have evidence, and yet no individual member of the group has the same evidence. For example, in double blind trials it may not be appropriate to attribute any of the individual scientists with possessing the evidence that, say, the drug is safe (since none of them would be justified in believing this), and yet it does seem appropriate to say that "there is evidence that the drug is safe" and that the most appropriate bearer of this evidence is the group itself (cf. Magnus 2007). It seems plausible that the same could be said for knowledge although it would be in need of argument. In cases where researchers, for example, collect data about different aspects of some phenomenon but are not aware of each other's results, and where this data is compiled and processed to generate new knowledge, it seems that there is knowledge even though no one believes it, although this does not demonstrate, per se, that it is the group that knows. Another option might be so-called "third-person epistemology" or Karl Popper's objective knowledge where it is more appropriate to say "it is known that p" than "S knows that p" (Stevenson 1999).

6. Clark (2010) and Adams and Aizawa (2010) provide useful summaries from each side of the fence.

7. Asch's experiments demonstrated that people tend to modify their own opinions to better match the opinion of the group, even when it differs greatly from theirs or contradicts common sense.

BIBLIOGRAPHY

Adams, Fredrick and Kenneth Aizawa. "Defending the Bounds of Cognition." In *The Extended Mind*, edited by Richard Menery, 67–80. Cambridge, MA: MIT Press, 2010.

Andrews, David. "Information Routeing Groups—Towards the Global Superbrain: Or How to Find Out What You Need to Know Rather Than What You Think You Need to Know." *Journal of Information Technology* 1, no. 1 (1986): 22–35.

Asch, Solomon E. "Effects of Group Pressure on the Modification and Distortion of Judgments." In *Groups, Leadership and Men*, edited by Harold Guetzkow, 177–90. Pittsburgh, PA: Carnegie Press, 1951.

Asch, Solomon E. "Studies of Independence and Conformity: A Minority of One Against a Unanimous Majority." *Psychological Monographs*, 70, no. 9 (1956): 1–70.

Blomberg, Olle. "Socially Extended Intentions-in-Action." *Review of Philosophy and Psychology* 2, no. 2 (2011): 335–53.

Bratman, Michael E. "Shared Cooperative Activity." *The Philosophical Review* 101, no. 2 (1992): 327–41.

Bratman, Michael E. "Shared Intention." *Ethics* 104, no. 1 (1993): 97–113.

Carter, J. Adam, Jesper Kallestrup, S. Orestis Palermos, and Duncan Pritchard. "Varieties of Externalism." *Philosophical Issues* 24, no. 1 (2014): 63–109.

Castelfranchi, Cristiano. "Minds as Social Institutions." *Phenomenology and the Cognitive Sciences* 13, no. 1 (2014): 121–43.

Clark, Andy. *Supersizing the Mind: Embodiment, Action, and Cognitive Extension.* Oxford: Oxford University Press, 2008a.

Clark, Andy. "Pressing the Flesh: A Tension in the Study of the Embodied, Embedded Mind?" *Philosophy and Phenomenological Research* 76, no. 1 (2008b): 37–59.

Clark, Andy. "Coupling, Constitution and the Cognitive Kind: A Reply to Adams and Aizawa." In *The Extended Mind* edited by Richard Menery, 81–100. Cambridge, MA: MIT Press, 2010.

Clark, Andy and David Chalmers. "The Extended Mind." *Analysis* 58, no. 1 (1998): 7–19.

Giere, Ronald N. "Discussion Note: Distributed Cognition and Epistemic Cultures." *Philosophy of Science* 69, no. 4 (2002a): 637–44.

Giere, Ronald N. "Scientific Cognition as Distributed Cognition." In *The Cognitive Basis of Science*, edited Peter Carruthers, Stephen Stich, and Michael Siegal, 285–99. Cambridge: Cambridge University Press, 2002b.

Giere, Ronald N. "Distributed Cognition without Distributed Knowing." *Social Epistemology* 21, no. 3 (2007): 313–20.

Gilbert, Margaret P. *Social Facts*. London: Routledge, 1989.

Gilbert, Margaret P. "Sociality as a Philosophically Significant Category." *Journal of Social Philosophy* 25, no. 3 (1994): 5–25.

Gilbert, Margaret P. "Belief and Acceptance as Features of Groups." *Protosociology: An International Journal of Interdisciplinary Research* 16 (2002): 35–69.

Gilbert, Margaret P. "Collective Epistemology." *Episteme* 1, no. 2 (2004): 95–107.

Goldberg, Sanford. *Relying on Others: An Essay in Epistemology*. Oxford: Oxford University Press, 2010.

Goldman, Alvin. "Group Knowledge Versus Group Rationality: Two Approaches to Social Epistemology." *Episteme* 1, no. 1 (2004): 11–22.

Hakli, Raul. "On the Possibility of Group Knowledge Without Belief." *Social Epistemology* 21, no. 3 (2007): 249–66.

Hetherington, Stephen. "The Extended Knower." *Philosophical Explorations* 15, no. 2 (2012): 207–218.

Hutchins, Edwin. *Cognition in the Wild*. London: MIT Press, 1995.

Kelp, Christoph. "Extended Cognition and Robust Virtue Epistemology." *Erkenntnis* 78, no. 2 (2013): 245–52.

Kerr, Eric T. and Gelfert, Axel. "The 'Extendedness' of Scientific Evidence." *Philosophical Issues* 24, no. 1 (2014): 253–81.

Knobe, Joshua and Jesse J. Prinz. "Intuitions About Consciousness: Experimental Studies." *Phenomenology and the Cognitive Sciences* 7, no. 1 (2008): 67–83.

Lackey, Jennifer. "Group Knowledge Attributions." In *Knowledge Ascriptions*, edited by Jessica Brown and Mikkel Gerken, 243–69. Oxford: Oxford University Press, 2012.

Lahroodi, Reza. "Collective Epistemic Virtues." *Social Epistemology* 21, no. 3 (2007): 281–97.

List, Christian. "Group Knowledge and Group Rationality: A Judgment Aggregation Perspective." *Episteme* 2, no. 1 (2005): 25–38.

List, Christian and Philip Pettit. "Aggregating Sets of Judgments: An Impossibility Result." *Economics and Philosophy* 18, no. 1 (2002): 89–110.

List, Christian and Philip Pettit. "Aggregating Sets of Judgments: Two Impossibility Results Compared." *Synthese* 140, no. 1–2 (2004): 207–35.

List, Christian and Philip Pettit. "Group Agency and Supervenience." *Southern Journal of Philosophy* 44, no. S1 (2006): 85–105.

Magnus, Paul D. "Distributed Cognition and the Task of Science." *Social Studies of Science* 37, no. 2 (2007): 297–310.

Maravita, Angelo and Atsushi Iriki. "Tools for the Body (Schema)." *Trends in Cognitive Sciences* 8, no. 2 (2004): 79–86.

Mathiesen, Kay. "The Epistemic Features of Group Belief." *Episteme* 2, no. 3 (2006): 161–75.

Merleau-Ponty, Maurice. *Phénomènologie de la Perception*. 1945. Translated by Colin Smith. *Phenomenology of Perception*. London: Routledge, 2002.

Palermos, Orestis and Duncan Pritchard. "Extended Knowledge and Social Epistemology." *Social Epistemology Review and Reply Collective* 2, no. 8 (2013): 105–20.

Pauly, Marc and Martin Van Hees. "Logical Constraints on Judgment Aggregation." *Journal of Philosophical Logic* 35, no. 6 (2006): 569–85.

Pettit, Philip. "Groups with Minds of Their Own." In *Socializing Metaphysics*, edited by Fred Schmitt, 167–94. Oxford: Rowman and Littlefield, 2003.

Pritchard, Duncan. "Cognitive Ability and the Extended Cognition Thesis." *Synthese* 175, no. 1 (2010): 133–51.

Quinton, Anthony. "Social Objects." *Proceedings of the Aristotelian Society* 75 (1975/1976): 1–27.

Rupert, Robert. "Minding One"s Cognitive Systems: When Does a Group of Minds Constitute a Single Cognitive Unit." *Episteme* 1, no. 3 (2005): 177–88.

Schmitt, Fredrick F. "The Justification of Group Beliefs." In *Socializing Epistemology: The Social Dimensions of Knowledge*, edited by F. F. Schmitt, 257–87. Lanham, MD: Rowman and Littlefield, 1994.

Seemann, Axel. "Joint Agency: Intersubjectivity, Sense of Control, and the Feeling of Trust." *Inquiry* 52, no. 5 (2009): 500–15.

Stevenson, Leslie. "First Person Epistemology." *Philosophy* 74, no. 4 (1999): 475–97.

Theiner, Georg. "Making Sense of Group Cognition." *ASC09: Proceedings of the 9th Conference of the Australasian Society for Cognitive Science*, edited by Wayne Christensen, Elizabeth Schier, and John Sutton. Sydney: Macquarie Centre for Cognitive Science, 2009.

Theiner, Georg and Timothy O'Conner. "The Emergence of Group Cognition." In *Emergence in Science and Philosophy*, edited by Antonella Corradini and Timothy O'Connor, 78–119. New York: Routledge, 2010.

Thomas, Nigel J. T. "Mental Imagery." In *The Stanford Encyclopedia of Philosophy*, edited Edward N. Zalta. Fall 2013. plato.stanford.edu/archives/fall2013/entries/mental-imagery/.

Tollefsen, Deborah. "Organizations as True Believers." *Journal of Social Philosophy* 33, no. 3 (2002): 395–410.

Tollefsen, Deborah. "Group Testimony." *Social Epistemology* 21, no. 3 (2007): 299–311.

Tumollini, Luca and Cristiano Castelfranchi. "The Cognitive and Behavioral Mediation of Institutions: Towards an Account of Institutional Actions." *Cognitive Systems Research* 7 no. 2–3 (2006): 307–23.

Tuomela, Raimo. "Group Knowledge Analyzed." *Episteme* 1, no. 2 (2004): 109–27.

Ward, Dave and Mog Stapleton. "Es Are Good: Cognition as Enacted, Embodied, Embedded, Affective and Extended." In *Consciousness in Interaction: The Role of the Natural and Social Context in Shaping Consciousness*, edited by Fabio Paglieri, 89–104. Amsterdam, John Benjamins Publishing Company, 2012.

Wilson, Robert A. *Boundaries of the Mind: The Individual in the Fragile Sciences*. Cambridge: Cambridge University Press, 2004.

Wilson, Robert A. "Extended Vision." In *Perception, Action and Consciousness*, edited by Nivedita Gangopadhyay, Michael Madary, and Finn Spicer, 277–89. New York: Oxford University Press, 2010.

Vaesen, Krist. "Knowledge Without Credit, Exhibit 4: Extended Cognition." *Synthese* 181, no. 3 (2011): 515–29.

Chapter 14

Disagreement and the Ethics of Belief

Jonathan Matheson

The question at the center of the debate concerning the epistemic significance of disagreement is the question how evidence that another disagrees with you affects the justificatory status of your beliefs (if at all)? Conciliatory views maintain that such evidence is a defeater for the justification that you have for your disputed beliefs.[1] If undefeated, such a defeater renders your disputed beliefs less justified, if not entirely unjustified. Strong conciliatory views, such as the Equal Weight View (EWV), appear to have the consequence that none of our controversial beliefs are epistemically justified (see Carey and Matheson 2013). This consequence of EWV is perhaps best seen by looking at an idealized two-person case of peer disagreement.[2] If two peers disagree about a proposition, say one believes it and the other disbelieves it, the disagreement is not better explained by positing an error with either one of the peers over and above the other. Since the disagreement is equally well explained by one party having made an error as by the other party having made an error, EWV claims that both parties are rationally mandated to revise their opinions to an attitude that "splits the difference" between the two original attitudes—in this case suspension of judgment. It is worth noting that having followed the prescriptions of EWV, the original disagreement disappears—having split the difference, both parties now adopt the very same doxastic attitude toward the originally disputed proposition.

However, the idea that we should abandon controversial beliefs and eliminate the disagreements surrounding them is challenged by psychological evidence about group inquiry.[3] In brief, there is evidence that groups with members who genuinely disagree about a proposition, do better with respect to determining whether that proposition is true. This evidence indicates that the elimination of disagreements is bad—and not just any kind of bad, it is *epistemically* bad. So, the prescriptions of EWV appear to be in tension with psychological findings regarding group inquiry.

In what follows, I will explain this challenge to EWV and evaluate its merits. We will see that while this evidence does not give us a reason to reject EWV, it does require making some clarifications regarding what the view does and does not entail, as well as a revisiting of the ethics of belief.

THE PSYCHOLOGY OF GROUP INQUIRY

Several related theses have considerable empirical support. I will only briefly lay them out here since my primary goal is not to evaluate the legitimacy of these findings, but rather what import they have on the epistemology of disagreement and questions concerning what we ought to believe.

> A. Groups reach more accurate conclusions when there is dissent and debate within the group.

First, groups do better than individuals in performing reasoning tasks. Some evidence for this is given by Moshman and Geil (1998). In their study, participants were asked to solve the Wason Selection Task in one of three ways: individually, interactively, or individually and interactively. Subjects who solved the task individually only relied on their own reasoning. Subjects who solved the task interactively encountered the task as a group and worked through it together. Subjects who solved the task individually and interactively first relied on their own reasoning to come to a conclusion, and then discussed and debated their individual findings in a group before coming to a final conclusion. The success rates for the various groups were 9.4 percent, 70 percent, and 80 percent, respectively. This indicates that reasoning together is more fruitful.

Further evidence suggests that debate and not merely discussion increases the accuracy of group inquiry. Sunstein (2002) provides evidence that like-minded groups tend to increase their confidence in their shared views, and that group discussion in such contexts only tends to further establish the held view, what Sunstein calls the "law of group polarization." So, there are also epistemic dangers that are increased with like-minded groups.

> B. Groups reach more accurate conclusions when dissenting parties genuinely hold their dissenting beliefs in the debate.

Evidence for B comes from Mercier (2011) who suggests that we tend to seek out arguments that confirm our antecedently held beliefs. Such confirmation bias is a detriment to both individuals and like-minded groups. In addition, Dawson et al. (2002) provide evidence that we are better at finding falsifying data regarding arguments and opinions with which we disagree. So, groups with genuinely dissenting parties will do better in terms of placing appropriate checks and balances on the reasoning and arguments deployed in the group. This conclusion is confirmed with work by Schulz-Hardt et al. (2002). They found that groups with genuine disagreement performed better than those where the disagreement was merely contrived by using various "devil's advocate" techniques. Evidence suggests that we reason better with respect to beliefs that we actually have than we do with respect to "devil's advocate" beliefs. So, groups with genuine dissent do better at avoiding both belief bias and confirmation bias.[4] So, diversity in opinion among members of a group helps to both uncover a more balanced body of evidence as well as to properly evaluate that evidence. Maintaining (at least some degree of) disagreement is thus quite beneficial, even epistemically speaking.

SHARPENING THE PROBLEM

The ethics of belief concerns the central question what should I believe? However, this question plausibly has many senses. One issue within the ethics of belief is whether this question is univocal or multiply ambiguous. After all, our beliefs have consequences in a number of areas of our lives, and in the lives of others. So, we can at least identify several more specific questions in the neighborhood, such as:

- From the moral point of view, what should I believe?
- From the prudential point of view, what should I believe?
- From the epistemic point of view, what should I believe?

So, an issue in the ethics of belief is whether there is one central unified question "what should I believe?" that comes from the *all-things-considered* perspective.[5] However, regardless of whether there is such a perspective, it should be clear that there are versions of this question from these more fine-grained perspectives. Of interest to us here is this question when asked from the epistemic perspective. It doesn't matter here whether (or how) an answer to this question informs an answer to the questions from the all-things-considered perspective.

Two goals are apparent from the epistemic perspective: to believe truths and to disbelieve falsehoods. Each of these epistemic goals holds the other in check, preventing the epistemic maladies brought about by either an overly cautious or an overly careless doxastic life. Maximizing one's chances of success with respect to one of these goals comes only by way of subjecting oneself to grave risk with respect to the other goal. So, while being an indiscriminate believer may increase the likelihood of my having true beliefs, such a strategy also makes it likely that I will have many false beliefs. Similarly, while refusing to have any beliefs at all (if even possible) would avoid even the potential for a false belief, it would also thwart my having any true beliefs. So, we can more clearly formulate the epistemic version of our question as *with the interest of believing truths and disbelieving falsehoods, what should I believe*?[6]

With this formulation of the epistemic perspective, we can see that both EWV and the recommendations coming from the psychology of group inquiry are concerned with our doxastic lives from the epistemic perspective. Each is concerned with the goals of believing truths and disbelieving falsehoods. For each, the goal is to arrive at true conclusions and to avoid false ones. Further, each is concerned with what I should believe *right now*, with respect to these epistemic goals. Neither is concerned with beliefs I may or may not have had in the past, and neither makes prescriptions for what I should believe in the future. Further still, each is concerned with what I should believe now *regarding the disputed proposition in question*, and with respect to the goal of believing it if it is true and disbelieving it if it is false. Each has the same target proposition in mind.

On the one hand, there is powerful evidence that abandoning one's dissenting belief upon discovering a disagreement has quite negative effects. Further, these negative effects are not psychological or aesthetic; they are *epistemic*. They are negative epistemic consequences since abandoning one's dissenting opinion actually hinders one's likelihood of having a true belief on the matter in the end. On the other hand, EWV claims that we should abandon controversial beliefs. This norm too is properly seen as epistemic—it is with the goal of believing truths and disbelieving falsehoods that it claims that we should be agnostic about the controversial. So, with respect to the epistemic goal of believing truths and disbelieving falsehoods, the prescriptions of EWV are in tension with the prescriptions of the psychological research. According to EWV, we ought to abandon our controversial beliefs, but according to the psychological data, we ought to (at least sometimes) keep them. We clearly cannot both keep them and abandon them, so what are we to do?

DISSOLVING THE PUZZLE

I think that it is plausible that our puzzle rests on a confusion and that this confusion arises by confusing two kinds of epistemic goods. These goods can be distinguished by differentiating between synchronic and diachronic epistemic norms or goals.

While we have identified the epistemic version of the question "what should I believe?" with the question "from the goal of believing truths and disbelieving falsehoods, what should I believe?" and while we have seen both EWV and the evidence from the psychology of group inquiry as providing answers to this very question, nevertheless, the respective epistemic motivations in question are distinct. However, we can account for the distinctiveness of these motivations while classifying both pursuits as epistemic in nature. I suggest that we do this by distinguishing between synchronic and diachronic epistemic norms, and thus by distinguishing synchronic and diachronic versions of the question "from the epistemic point of view, what should I believe?"

According to the synchronic version of this question, what matters is how well I succeed at believing truths and disbelieving falsehoods *right now*. According to the diachronic version of the question, what matters is how well I succeed at believing truths and disbelieving falsehoods *in the long run*. While they both concern what I should believe right now in terms of fulfilling these goals, and while they are both concerned with epistemic goals (the twin epistemic projects of believing truths and disbelieving falsehoods), the times of appraisal differ. That is, the synchronic and diachronic versions of the question differ with respect to when they see my epistemic goals as being achieved. The synchronic version concerns how well I am doing at my epistemic projects *right now*, while the diachronic version concerns how well I am doing right now with regard to my *long-term* epistemic goals. These questions differ in *where*, or better *when*, they locate the relevant success and failure conditions, even if they both offer doxastic prescriptions for the agent *right now*, and even if those prescriptions are both properly conceived of as epistemic.

Returning to the issue at hand, EWV is best seen as giving us an answer to the synchronic version of our question. It claims that abandoning controversial beliefs is the best way to be synchronically epistemically rational. The data coming from the psychology of group inquiry is best seen as giving an answer to the diachronic version of our question. It supports the claim that maintaining controversial beliefs is (at least sometimes) the best way to be diachronically epistemically rational. So, rather than motivating a prescription that conflicts with EWV, the evidence from the psychology of group inquiry motivates a distinct answer to a distinct question—a question that EWV simply offers no answer to. So, the psychological research is best seen as engaging in a different (though related) question than the question that EWV is addressing. The conciliatory answer to the synchronic question places no pressure on a steadfast answer to the diachronic question, and a steadfast answer to the diachronic question places no pressure on a conciliatory answer to the synchronic question.

A WORRY EMERGES

While the response on offer may resolve the initial puzzle, it can be thought to only make matters worse. After all, one of the things we may have thought we would get, and something we wanted, from EWV (and other views of the epistemic significance of disagreement) is a prescription regarding what we should believe, or minimally what we should believe from the epistemic perspective. However, if the question "from the epistemic perspective, what should I believe?" is ambiguous and EWV only offers an answer to one of the disambiguated questions (the synchronic version), then it simply fails to tell us what to believe (even from the epistemic perspective). It fails to give an answer to the *all-epistemic-things-considered* version of our question since it does not tell us how to weigh answers to the synchronic version of the question with the answers to the diachronic version of the question, and it is compatible with there being no way of combining these answers at all. That is, it may be that there simply is no all-epistemic-things-considered perspective, but only the disambiguated epistemic perspectives. This parallels the thought that there is no all-things-considered answer to the "what should I believe?" question, but only answers to the moral, prudential, and epistemic disambiguated questions.

So understood, EWV does not claim that the balance of epistemic reasons point to agnosticism about controversial propositions. Rather, it only claims that the balance of epistemic reasons *of a certain sort* (the synchronic reasons), points toward agnosticism about controversial propositions. It makes no claims about how significant this sort of epistemic reasons are, or how (if at all) they figure in to what one has *all-epistemic-reasons-considered* reason to believe or what one has *all-things-considered* reason to believe. So, on this view, EWV does not mandate agnosticism about the controversial, it does not even mandate agnosticism about the controversial from the epistemic perspective. Rather, it merely mandates agnosticism about the controversial from the synchronic epistemic perspective, but as we have seen there is more to the epistemic perspective than this.

Having paired down what EWV is up to, it is worth noting why this consequence of EWV is still significant. First, the synchronic version of our question is one that is intrinsically interesting. It is worth considering what we should believe, with the goals of believing truths and disbelieving falsehoods, right now. Even if this question was novel, it is a question worthy of our consideration—a question of the type philosophers should consider. Second, this question is not novel. Plausibly, this is the very question that we have been asking when we have asked what beliefs are epistemically justified. While "justification," "rational," and their cognates are far from univocal, it is plausible that (at least historically) epistemologists have been examining epistemic justification (at least when seen as what helps put us in a position to know at the time in question) and have been concerned with answering the synchronic version of our question.

While we can see that both versions of the question are properly epistemic, and both are concerned with what is epistemically rational or justified for me to believe now, the diachronic answer does not provide the type of epistemic justification required for knowledge—at least not knowledge at the time in question. While remaining steadfast in one's controversial belief may best lead to knowledge about the matter in the long run, it does not supply the type of epistemic justification required for knowledge at the earlier time. But to say this is not to say that there is not some real sense in which those steadfast beliefs are epistemically justified. I think that plausibly they are, it is just that this is not the sense of epistemic justification that traditional epistemologists have been concerned with. And to say that, is not to denigrate it in any way. To say that it is a separate question is not to say that it is a less interesting or less important question—it is simply a different question.

HOW SKEPTICAL IS EWV?

Given all of this, the skeptical implications of EWV must be reassessed. Exactly how skeptical is EWV? I have argued elsewhere that the skeptical ramifications of the view are quite vast, however, the considerations here may seem to undermine those arguments (see Carey and Matheson 2013). So long as we understand "skeptical consequences" as consequences that our beliefs lack the kind of justification required for knowledge (note it is the kind of justification here, not the amount of justification), EWV does have quite skeptical consequences. After all, according to EWV our controversial beliefs lack that kind of justification. However, if we understand "skeptical consequences" as consequences that our beliefs lack *all* kinds of epistemic justification or that they are not on balance supported from the all-epistemic-things-considered point of view, then EWV simply does not have skeptical consequences. Just as EWV did not address whether there are any moral, prudential, or religious reasons why we may hold on to some controversial beliefs it similarly did not take on any diachronic epistemic reasons why we may hold on to our controversial beliefs. So, if one was looking for a more wide-reaching prescription from EWV, it is simply not there to be found. The consequence that our controversial beliefs lack the kind of epistemic justification required for knowledge is still striking and significant, but we should not be so quick as to extrapolate from this to any conclusions about what the epistemic perspective requires or what the all-things-considered perspective requires.

NOTES

1. For defenses of Conciliatory Views, see Feldman (2006), Elga (2007), Christensen (2007), and Matheson (2009).

2. Two individuals are *epistemic peers* with respect to a proposition *p* at a time *t*, just in case both individuals are in an equally good epistemic position with respect to *p* at *t*—both are equally likely to be correct about *p* at *t*. According to EWV, peer opinions are to be given equal weight.

3. See Dunn (2013) for a careful posing of this problem.

4. For more on confirmation bias, see Nickerson (1998).

5. For reasons to doubt the coherence of this question, see Feldman (2000).

6. The representativeness of this view of the epistemological perspective can be seen in David's "Truth as Epistemic Goal." David provides an extensive list of quotations to this end including epistemologists such as Alston, BonJour, Chisholm, Descartes, Foley, Goldman, Lehrer, Moser, Plantinga, and Sosa. This is not to say that truth is the *only* thing of value from the epistemic perspective, only that everything that has value from the epistemic perspective has it due to its relation to truth. For a nice discussion of this and related issues, see Pritchard (2014).

BIBLIOGRAPHY

Carey, Brandon and Jonathan Matheson. "How Skeptical is the Equal Weight View?" In *Disagreement and Skepticism*, edited by Diego Machuca, 131–49. New York: Routledge, 2013.

Christensen, David. "Epistemology of Disagreement: The Good News." *Philosophical Review* 116 (2007): 187–218.

David, Marian. "Truth as the Epistemic Goal." In *Knowledge, Truth, and Duty: Essays on Epistemic Justification, Virtue, and Responsibility*, edited by Matthias Steup, 151–69. Oxford: Oxford University Press, 2001.

Dawson, Erica, Thomas Gilovich, and Dennis T. Regan. "Motivated Reasoning and Performance on the Wason Selection Task." *Personality and Social Psychology Bulletin* 28, no. 10 (2002): 1379–87.

Dunn, Jeffrey. "Peer Disagreement and Group Inquiry." Paper presented at annual meeting of the Indiana Philosophical Association, Indianapolis, Indiana, October 2013.

Elga, Adam. "Reflection and Disagreement." *Noûs* 41 (2007): 478–502.

Feldman, Richard. "The Ethics of Belief." *Philosophy and Phenomenological Research* 60 (2000): 667–95.

Feldman, Richard. "Reasonable Religious Disagreements." In *Philosophers without Gods: Meditations on Atheism and the Secular Life*, edited by Louise Antony, 194–214. New York: Oxford University Press, 2006.

Matheson, Jonathan. "Conciliatory Views of Disagreement and Higher-Order Evidence." *Episteme: A Journal of Social Philosophy* 6 (2009): 269–79.

Mercier, Hugo. "Reasoning Serves Argumentation in Children." *Cognitive Development* 26 (2011): 177–91.

Moshman, David and Molly Geil. "Collaborative reasoning: Evidence for collective rationality." *Educational Psychology Papers and Publications* 52 (1998): 231–48.

Nickerson, Raymond. "Confirmation Bias: A Ubiquitous Phenomenon in Many Guises." *Review of General Psychology* 2 (1998): 175–220.

Pritchard, Duncan. "Truth as the Fundamental Epistemic Goal." In *The Ethics of Belief: Individual and Social*, edited by Jonathan Matheson and Rico Vitz, 112–29. New York: Oxford University Press, 2014.

Schulz-Hardt, S., M. Jochims, and D. Frey. "Productive Conflict in Group Decision Making: Genuine and Contrived Dissent as Strategies to Counteract Biased Information Seeking." *Organizational Behavior and Human Decision Processes* 88 (2002): 563–86.

Sunstein, Cass. "The Law of Group Polarization." *Journal of Political Philosophy* 10 (2002): 175–95.

Chapter 15

Doxastic Involuntarism, Attentional Voluntarism, and Social Epistemology

Mark Douglas West

In a Mertonian sense, linkage issues (Hedstrom and Udehn 2009) connecting lower-range hypothetical and empirical constructs to the larger-scale psychological constructs of our domain are still nascent. Social epistemology, from its inception, represents a daring and synthetic break with the traditional and stifling boundaries of science and philosophy (Fuller and Collier 2003); now that that break has been made, the work of fully conceptualizing our theories (Glaser 2002) and "grounding" our theories in empirical research has begun.

To fully conceptualize our theories would mean, as Merton (1967) suggests, that we fully articulate the linkages between the grand theories at the highest conceptual levels of abstraction and the empirical tests at the lowest levels of abstraction in our mental explanatory schemata. To "ground" a middle-range theory (Boudon 1991) such as what I take the theory of social epistemology to be in practice requires empirical tests of hypotheses derived from that theory, as well.

I propose some of those ideas here. I argue that beliefs are a special sort of feeling about the truth value of statements. Once that conclusion is drawn, beliefs can be seen to have little to do with what is really of import in epistemology—the formation of shared meanings. I then argue that doxastic involuntarism suggests that we must examine something else—attentional voluntarism—if we are to understand how agents change behaviors, and that once we examine attentional voluntarism, we are thrust into the social realm, and into social epistemology. Throughout, I will be discussing Alcoholics Anonymous, an organization whose use of symbolic language and social support has been demonstrated to create a social epistemology for its members which has an empirical effect upon their behavior. By so doing, I hope to suggest a sort of praxis by which we might "ground" social epistemology, which I regard as a grand theory in the Mertonian sense.

BELIEFS

What are beliefs? That question, while simple enough to formulate, is quite difficult to answer. And a good deal hinges upon its answer. The dominant faiths in the world in the twenty-first century—Christianity and Islam, which together claim 54.7 percent of the world's population as adherents (Hackett 2012)—demand that their followers "believe" their tenets willingly, and, pursuant to that belief, confess them to others. The awareness that much of the philosophical world has turned its back on doxastic voluntarism as an untenable position, while Christianity in particular holds such voluntaristic concepts of faith as a centerpiece of what it means to be an adherent of their religion, has resulted in significant tension within the theological hierarchy of that religion (Hartman 2011).

We might see the contrast between the epistemic views of a small, highly educated group of individuals—philosophers and theologians—and those of a much larger group of less-educated group of individuals—the laity—as evidence of epistemic communities (Cross 2013). These communities exist in different locations, with the doxastic involuntarists centered in urban areas around institutions of higher learning and the doxastic voluntarists scattered around the globe.

Taken as a whole, such data supports two arguments. First, social formations are crucial to the epistemic venture, and can be employed in a Comtean manner in the reconstitution of the epistemic domain a la Fuller (1987); second, these social formations have socioeconomic differences that can be used as data in empirical models that seek to predict epistemological differences a la Vähämaa and West (2014).

VOLUNTARISM AND INVOLUNTARISM

The question of whether beliefs may or may not be justified is in itself troubling enough—see, for example, Oakley (1976) for the argument that it may be impossible to justify beliefs at all. More important, however, is the fact the idea that we can choose what to think and believe, which in at least some significant sense lies at the heart of the modern notion of the autonomous rational agent capable of self-determination (Schneewind 1991). This conceptualization of the self is important both to democratic notions and to psychological notions; the debates surrounding rationalism and voluntarism at the very dawn of the Protestant revolution (Schneewind 1996) have to do with *de servo arbitrio*, as Luther had it. As Schneewind and others (Erdelack 2011) have suggested, such questions served as the fuel for later debates about the most important questions of ethics and volition in the modern age.

The debates for doxastic voluntarism and involuntarism proceed by analogy (Nottelmann 2006). A question posed by Alston (1989, 122) is operative: "I shall merely contend that we are not so constituted as to be able to take up propositional attitudes at will. My argument for this, if it can be called that, simply consists in asking you to consider whether you have any such powers."

Imagine that I asked you to imagine that I was General George Custer, and offered you $10,000 USD to do so. You might squint, puzzle upon the thesis, and then assert that you had succeeded. You would then address me as "George," ask me how things were going in my preparations for Little Big Horn, comment upon my lush golden hair, and so on. But could you truly say that you could alter your beliefs? You could alter your behaviors to be sure, but your *beliefs* would remain unaltered. I would surmise, and rightly so, that your actions with regard to addressing me as "George" were instrumental, based on the hope of persuading me that your beliefs had in fact changed so that the $10,000 USD would become yours. A more extreme argument from Booth (2007)—can you, for a very large sum of money you desperately need, imagine me to be a grasshopper—has a similarly persuasive outcome.

These arguments, for the majority of the philosophical "in-groups" described earlier, have proven persuasive against doxastic voluntarism. But those supporting doxastic voluntarism have a counterargument. How, then, is it that people change? Witness the success of Alcoholics Anonymous in the remission of alcoholism (Krentzman et al. 2011); that organization holds that remission from alcoholism begins with a first step:

> *We admitted we were powerless over alcohol—that our lives had become unmanageable.*

Such an admission seems to indicate "taking up a propositional attitude at will"; as the old joke concerning psychotherapy has it, "the light bulb has to want to change." Alcoholics Anonymous is a particularly useful example in that the organization represents not only an organization dedicated to a volitional change of a belief (its members go from thinking that the consumption of alcohol is a pleasant and worthwhile action, to something to be avoided at all costs). Alcoholics Anonymous has been empirically studied and shown to have significant efficacy as a modality for treatment for alcohol addiction (Roman 1988).

Despite such counterarguments, doxastic involuntarism has largely held sway in the philosophical realm with Heil (1983), Kornblith (1982), and Price (1954) presenting standard arguments. Another argument for doxastic voluntarism, one that suggests self-deception and magical thinking are ways in which individuals volitionally trump evidence, is dealt with by

Cote-Bouchard (2012, 14–15). In essence, Booth (2007) advances a similar argument; self-deception is psychologically impossible. How, exactly, would an agent contrive to deceive himself or herself successfully regarding a concept, such that they would come to believe some proposition S that was false? How, exactly, would an agent forget the act of self-deceit if it were truly volitional?

Nevertheless, support remains (Holyer 1983; Govier 1976; O'Hear 1972) for doxastic voluntarism. The arguments for voluntarism, as suggested above, arises from a persistently felt sense that humans have free will; if we are free to change, then we must surely be free to believe at least some things, at least some things about agency, in order for that change to occur.

THE DILEMMA

Hence, as we mentally examine the analogies presented by doxastic involuntarists and voluntarists, we are faced with a conundrum. We clearly cannot force ourselves to believe anything we like; yet people change. Further, we know how they change; they announce that they want to change; then they cast about for how they might change, and eventually they land upon some behavioral mechanism that works, and they change.

Our "felt sense" of the world, then, is that we don't choose what we believe, or, really, think. We look at evidence, and from there, "arrive" at conclusions. We may "feel" that we have chosen those conclusions, but the evidence has, so to speak, "forced our hands." We could not reasonably see, as do all observers, that having two things, and getting two more, results in us having four things, and then conclude from that, contrary to all other observers, that two plus two equals five; the evidence compels us to conclude that two plus two equals four. The "felt sense" we have is a matter of no import; to conclude otherwise, as Plato (1976) has Socrates suggest in the Protagoras, is perverse.

On the other hand, agents *do* change their behaviors in a purposeful manner, and that leads us to have the feeling that we have control over our lives. We decide, at some point, that we weigh too much, or that smoking tobacco harms our health, and we modify our behavior. We know that we have volition, and hence free will; but our "felt sense" in this case collides with our "felt sense" in the case of the beliefs that impelled us to take up the cause of change in the first place. Most books on the subject, whether intended for scholarly (Goldstein and Kanfer 1975) or popular (Wheelis 1973) audiences, sensibly skirt the issue of exactly how people come to believe that they need to change and instead focus on what should or does happen from that point onward.

The question remains, however. How do we change our behavior, if we are unable to volitionally change our beliefs? As Fishbein and Ajzen (1975) suggest, beliefs give rise to behaviors, and hence doxastic involuntarism suggests determinism. Yet, we are dispositionally committed to freedom of the will. How do we resolve this conundrum?

ATTENTIONAL VOLUNTARISM

We may think ourselves out of the above dilemma by way of attentional voluntarism. Clarke (1986) offers an important clue about how people actually engineer change. The individuals who attend Alcoholics Anonymous, say, have indeed come to believe that "they were powerless over alcohol" by dint of perceptions that came their way without (or against) their will. But notice that their next steps are social. They affiliate themselves with an Alcoholics Anonymous group, and they stop going to local bars. They find a sponsor within the movement. They attend meetings, in which they receive social support of various sorts.

Further, they learn a new set of symbols and meanings related to alcohol, in which the concept of "poison" predominates—alcohol poisons relationships and the body (Antze 1987), alcohol as (metaphorical) storm or fire, destroys property and health (Jensen 2000). The old reinforcements of the pleasures of drink and of conviviality at bars are gone, and are replaced by the camaraderie of Alcoholics Anonymous. Via the new affective metaphors taught in the group, entrained through a shift in attention arising through the voluntary creation of new social links, the individual is able to begin the process of remission from alcohol addiction.

As Clarke (1986, 43) says:

> By ignoring the adverse evidence, searching for positive evidence, and concentrating on a particular reinforcing proposition, one has influenced one's belief acquisition processes. One has put oneself in a position not to receive evidence that would force one to believe something one does not want to believe. To this extent, and in this respect, "attention voluntarism" is a true doctrine.

The nascent alcoholic has sensory inputs that tell him or her that drinking is pleasant, and so he or she drinks. Later, they have become addicts, and they have sensory inputs that lead them to develop feeling states that move them to new behaviors concerning alcohol—in this case, the turn to new social structures, within which they adopt a new epistemology. This new epistemology entails a new way of understanding their world and the things in it, as detailed in *The Blue Book* and other literature from Alcoholics Anonymous. Alcohol itself takes on the role of a boundary object (Knorr-Cetina 1999; Trompette and Vinck 2009), an object that maintains a common identity across communities but serves to highlight the interpretive differences between those communities. Such groups, to use a phrase from Vähämaa (2013, 3), are epistemic communities—groups whose goals are diverse, but in which individuals adopt modes of thought that define knowledge in particularlistic ways. As Roche (1989) argues, the goal of personal action is the maximization of personal well-being; Vähämaa extends that Aristotelian line of reasoning to include the agent in an epistemological context. In the case of the alcohol addict, the goal of group membership is freedom from alcohol addiction, not a true understanding of the etiology of the nature of their addiction. The symbolic content of the message of Alcoholics Anonymous ("alcohol as poison," in several different registers) may or may not be "true" in some veritistic sense (Goldman 1999); its value to the agent is in its efficacy in reducing suffering. "Truth," as well as "sobriety," for the member of Alcoholics Anonymous, is redefined as honesty (Kurtz 1979, xii), and the epistemological dimension of the organization can be shown through its redefinition of critical terms in its venture to reshape behavior.

WHAT IS BELIEF?

Alcoholics Anonymous speaks of "belief" primarily in terms of Step 2 of the Twelve Steps, in which adherents are asked to turn themselves over to a "Higher Power." While belief in a "higher power" is widely understood in Alcoholics Anonymous to be essential to success in symptom relief, Murray et al. (2003) found that belief in an internal locus of control with regard to alcohol avoidance predicted a greater level of success with cessation than did belief in an external locus of control. In the context of Alcoholics Anonymous, then, "belief" as it is commonly construed has little meaning in the sense of faith.

Such a finding highlights the difficulty of talking about belief. This problematic has been explored in a number of contexts (Lindquist and Coleman 2008; Netland 1986; Gardenfors 1990; de Lavalette and Zwart 2011). Simply put, "belief" and "to believe" are in general terms used loosely, even in philosophical discourse. As Smith (1994, 21) argues, however, there is at least a general concurrence that there is some connection between "belief" and "desire"—where there is a belief, there is at least a desire that the statement involved in the belief be, or seek to be, true; belief, at the very least, involves a statement, which one asserts to be to the best of one's knowledge, true.

But even this minimal definition is problematic. Let's consider statements such as those Winch (1996) proposes, which derives from Wittgenstein's verbal formulation of Moore's Paradox in the Philosophical Investigations, where individuals contradict themselves, as in a statement like "I cannot believe I am standing here before you tonight, and yet I am." As Winch suggests, Wittgenstein goes on to suggest that all assertions have a human as their author, and all (rational) statements are assertions of truth (Wittgenstein 1974, Tractatus 5.5422): "The correct explanation of the form of the proposition 'A judges p' must show that it is impossible to judge a nonsense."

Note that Wittgenstein uses "judges" and "thinks" and "asserts" interchangeably, and, as Winch argues is the position of Wittgenstein, all assertions are implicitly predicated with "I believe that."

Thus, I argue, as Wittgenstein seems to suggest, that belief is an attitude toward a proposition. Propositions are sentences (Gardenfors 1990) whose informational content *seems* to be about a state of affairs in the world, but which is *in fact* referring to the state of mind of the individual making the utterance. Thus, the statement referred to earlier would expand to "I cannot believe I am standing here before you tonight, and yet I *believe* I am," and the implied believe makes the sentence self-contradictory, or at least referential to an incoherent state of mind.

Further, as Liska (1984, 62) argues, following Fishbein and Ajzen (1975; Ajzen and Fishbein 1980), an attitude is thus an affective evaluation of a proposition.

It (the Fishbein and Ajzen model) conceptually distinguishes the three components of the traditional attitude concept (affect, cognition, and conation) and specifies a recursive-chain causal structure underlying them. It assumes that behavior is directly caused by behavioral intentions (conation), which are caused by attitudes (affective evaluations), which in turn reflect beliefs about the consequences of behavior weighted by the subjective evaluation of the consequences. (Fishbein and Ajzen use the term "attitude" to refer to the affective evaluative dimension.)

Thus, a belief is a feeling (an affective evaluation) toward a cognition (a proposition, expressed as a sentence). The utterer may tell the truth, may lie, may be incorrect, or some combination of the above, but a belief is a feeling. And, as Mele (1989) suggests, it is entirely possible to have akratic *feelings*, feelings one has against one's better judgment. A nervous speaker may tell himself or herself of the benefits to be had by giving a relaxed speech, but the feelings persist. The speaker has applied an appropriate cognitive remedy, but he or she is still anxious; his or her emotional state is hence akratic. The involuntary state of those feelings is manifest. Just as one may have akratic beliefs, one may have akratic feelings. Indeed, I would argue that a feeling is indeed a feeling is an affective evaluation toward a belief.

At no point in the Fishbein and Ajzen model (perceptions to attitudes to behavioral intentions to behaviors, under the sway of perceived subjective norms) do we see the apparent necessity of the influence of volition. What does become manifest is the sufficiency of evaluative states to form behaviors; evaluative states have a good deal of consistency over time (Ledgerwood et al. 2010). Individuals can have a general disposition to dispositional evaluative states (Hepler and Albarracin 2013), and individuals with fewer numbers of affective evaluations (feelings) on a given topic are more likely to form their behaviors on the basis of social cues or pressures than are those with higher numbers of affective evaluations (Ledgerwood and Callahan 2012).

This last empirical finding is important for the social epistemologist. If affective evaluations are the source of behavioral intentions, and in turn, behaviors, as Fishbein and Ajzen (1975) argue and later empirical research suggests, then it should not be the case that fewer feelings lead to different sorts of behavior in the case of those with less direct experience of some phenomenon (who then turn to the social realm). What must be the case is that they turn to other feelings, and those other feelings are feelings that arise as a result of direct or indirect (mediated) social experiences.

EPISTEMOLOGY AND GROUP MEMBERSHIP

Do individuals, then, turn to groups in order to alter their epistemologies? In the case of Alcoholics Anonymous, almost certainly not. People do not think in those terms; no one ever says they "have a faulty epistemology, and need a better one." People, rather, turn to groups like Alcoholics Anonymous in order to be free from addiction—and changes in manners of thought are the modality by which they find relief. "Stinkin' thinkin'," the term Alcoholics Anonymous uses for the thought patterns associated with alcoholism, but not necessarily with drinking (Gorski and Miller 1982) is cured by "truth," which Alcoholics Anonymous construes as honesty. "Honesty," in turn, is verisimilitude in interpersonal relations, and adherence to a set of meanings assigned to concepts by the group; it is demonstrated by a set of behaviors that lead to sobriety. The group is the provider of a social epistemology, whose adoption leads to the behaviors that lead to a desired state—the cessation of a given type of suffering.

Alcoholics Anonymous is a very specific sort of group, and, of course, not all groups are so goal directed. But, as Vähämaa (2013) suggests, individuals join groups for instrumental reasons which all, ultimately, involve the desire for well-being, or eudaimonia. Groups provide social knowledge and cues to behavior, and that group knowledge is at least in part the linkage to the eudaimonia that groups provide. The turn to a group involves a desire to change behavior; the desire to change epistemologies, or ways of knowing, need never appear. The alcoholic wants to stop drinking, not to learn a new mode of knowing; yet they come to have both.

The same is true for the individual who joins a political party. One might, in the U.S. context, join the Republican Party hoping to gain useful acquaintances for business purposes, or to make friends, or to become more informed about local politics. What might well happen is that one might become persuaded that, say, science matters less than politics in the determination of how best to handle global warming (Demeritt 2001; Vähämaa 2013). The same would hold true for almost any social group; groups have group epistemologies and group membership eventually entails the adoption of a group epistemology.

CONCLUSION

The social is the domain in which humans operate. We turn to groups because we are social beings; being outside of groups has empirically demonstrable health consequences (Hawkley et al. 2003; Cacioppo et al. 2006), and, in turn, membership in voluntaristic groups have profound effects upon socialization (Harris 1995), the political process (Langton 1967), and a whole host of other crucial domains, in the adult as well as the child (Mortimer and Simmons 1978). Such socialization makes society possible, and forms the matrix in which "the social" exists.

Plato (1976, 345d–e) has Socrates argue in the Protagoras that *akrasia* is impossible, he suggests that one can only act in accord with what one knows, and that agents acted in accord with what they thought to be the good. But this is an ideal; our knowledge is always limited, and so we turn to "folk epistemologies" (Mercer 2008; Mercier 2010), homemade tests of credibility or likability (Kellner 1993) and, at a less-obvious level, various mechanisms of selective perception (Johnston and Dark 1986) and selective retention (Surlin and Gordon 1976), both in the psychological and the Campbellian sense (Simonton 2010).

At every turn, however, these heuristics are cognitively expensive, and the movement to the social has numerous benefits beyond the avoidance of *akrasia*. The ancients understood that humans were embedded in the social, and that with limited knowledge, we each sought to act in accord with what the groups we were embedded within had led us to understand was correct. It is only in a modern age, in which a theory of the self as "Victor and Invictus" (Weinstock 1957), the sovereign master of itself under all circumstances, has emerged. Such an autonomous rational agent, *Homo oeconomicus*, controls its thoughts, its actions, and its beliefs; it owes nothing to agents outside its own mind. Such reductionist strategies can be seen as impoverishing to the arts and interpersonal relations (Glynn 2005), privilege the current economic and political system (Read 2009), and as removing the social dimension from consideration in many domains *tout court*. I would argue that such a definition of the human has also limited our understanding of the nature of knowledge itself; and it is through a return to the social that we can work our way out of some of the problematics that trouble us in the realm of epistemology.

BIBLIOGRAPHY

Ajzen, Icek and Martin Fishbein. *Understanding Attitudes and Predicting Social Behavior.* Englewood Cliffs, NJ: Prentice-Hall, 1980.

Alston, William P. *Epistemic Justification: Essays in the Theory of Knowledge.* Ithaca: Cornell University Press, 1989.

Antze, Paul. "Constructive Drinking." In *Constructive Drinking: Perspectives on Drink from Anthropology*, edited by Mary Douglas, 149–81. Cambridge: Cambridge University Press, 1987.

Booth, Anthony R. "Doxastic Voluntarism and Self-Deception." *Disputatio* 2, no. 22 (2007): 115–30.

Boudon, Raymond. "What Middle-Range Theories Are." *Contemporary Sociology* 20, no. 4 (1991): 519–22.

Cacioppo, John T., Mary E. Hughes, Linda J. Waite, Louise C. Hawkley, and Ronald A. Thisted. "Loneliness as a Specific Risk Factor for Depressive Symptoms: Cross-Sectional and Longitudinal Analyses." *Psychology and Aging* 21, no. 1 (2006): 140–51.

Clarke, Murray. "Doxastic Voluntarism and Forced Belief." *Philosophical Studies* 50, no. 1 (1986): 39–51.

Cote-Bouchard, Charles. "Peut-on Etre Blame Pour Ses Croyances?: Le Deontologisme Epistemique Face au Probleme de L'involontarisme Doxastique." *These de Doctorat non Publiee*, Universite de Montreal, 2012.

Cross, Mai'a K. Davis. "Rethinking Epistemic Communities Twenty Years Later." *Review of International Studies* 39, no. 1 (2013): 137–60.

de Lavalette, Gerard R. and Sjoerd D. Zwart. "Belief Revision and Verisimilitude Based on Preference and Truth Orderings." *Erkenntnis* 75, no. 2 (2011): 237–54.

Demeritt, David. "The Construction of Global Warming and the Politics of Science." *Annals of the Association of American Geographers* 91, no. 2 (2001): 307–37.

Erdelack, Wesley. "Antivoluntarism and the Birth of Autonomy." *The Journal of Religious Ethics* 39, no. 4 (2011): 651–79.

Fishbein, Martin and Icek Ajzen. *Belief, Attitude, Intention, and Behavior: An Introduction to Theory and Research.* Reading, MA: Addison-Wesley, 1975.

Fuller, Steve. "On Regulating What is Known: A Way to Social Epistemology." *Synthese* 73, no. 1 (1987): 145–83.

Fuller, Steve and James H. Collier. *Philosophy, Rhetoric, and the End of Knowledge: A New Beginning for Science and Technology Studies.* Oxford: Routledge, 2003.

Gardenfors, Peter. "Belief Revision and Relevance." *PSA: Proceedings of the Biennial Meeting of the Philosophy of Science Association, Volume Two: Symposia and Invited Papers*, 1990, 349–65.

Glaser, Barney G. "Conceptualization: On Theory and Theorizing Using Grounded Theory." *International Journal of Qualitative Methods* 1, no. 2 (2002): 23–38.

Glynn, Simon. "The Atomistic Self Versus the Holistic Self in Structural Relation to the Other." *Human Studies* 28, no. 4 (2005): 363–74.

Goldman, Alvin I. *Knowledge in a Social World.* Oxford: Clarendon Press, 1999.

Goldstein, Arnold P. and Frederick H. Kanfer. *Helping People Change: A Textbook of Methods.* 4th ed. New York, NY: Pergamon Press, 1975.

Gorski, Terence T. and Merlene Miller. *Counseling for Relapse Prevention.* Independence, MO: Independence Press, 1982.

Govier, Trudy. "Belief, Values, and the Will." *Dialogue* 15, no. 4 (1976): 642–63.

Hackett, Conrad. "The Global Religious Landscape: A Report on the Size and Distribution on the World's Major Religious Groups as of 2010 (Rapport Techniquc)." Washington, DC: Pew Research Center. Pew-Templeton Global Religious Futures Project, 2012.

Harris, Judith R. "Where is the Child's Environment? A Group Socialization Theory of Development." *Psychological Review* 102, no. 3 (1995): 458–89.

Hartman, Robert J. "Involuntary Belief and the Command to Have Faith." *International Journal for Philosophy of Religion* 69, no. 3 (2011): 181–92.

Hawkley, Louise C., Mary Burleson, Gary Berntson, and John T. Cacioppo. "Loneliness in Everyday Life: Cardiovascular Activity, Psychosocial Context, and Health Behaviors." *Journal of Personality and Social Psychology* 85, no. 1 (2003): 105–20.

Hedstrom, Peter and Lars Udehn. "Analytical Sociology and Theories of the Middle Range." In *The Oxford Handbook of Analytical Sociology*, edited by Mary Douglas, 25–47. New York: Oxford University Press, 2009.

Heil, John. "Doxastic Agency." *Philosophical Studies* 43, no. 3 (1983): 355–64.

Hepler, Justin and Dolores Albarracin. "Attitudes Without Objects: Evidence for a Dispositional Attitude, its Measurement, and its Consequences." *Journal of Personality and Social Psychology* 104, no. 6 (2013): 1060.

Holyer, Robert. "Belief and Will Revisited." *Dialogue* 22, no. 2 (1983): 273–90.

Jensen, George H. *Storytelling in Alcoholics Anonymous: A Rhetorical Analysis.* Carbondale, IL: SIU Press, 2000.

Johnston, William A. and Veronica Dark. "Selective Attention." *Annual Review of Psychology* 37 (1986): 43–75.

Kellner, Hans. "Afterword: Reading Rhetorical Redescriptions." In *Rethinking the History of Rhetoric: Multidisciplinary Essays on the Rhetorical Tradition*, edited by Takis Poulakos, 241–56. Boulder, CO: Westview Press, 1993.

Knorr-Cetina, Karin. *Epistemic Cultures: How the Sciences Make Knowledge.* Cambridge, MA: Harvard University Press, 1999.

Kornblith, Hilary. "The Psychological Turn." *Australasian Journal of Philosophy* 60, no. 3 (1982): 238–53.

Krentzman, Amy R., Elizabeth Robinson, Brian E. Perron, and James A. Cranford. "Predictors of Membership in Alcoholics Anonymous in a Sample of Successfully Remitted Alcoholics." *Journal of Psychoactive Drugs* 43, no. 1 (2011): 20–26.

Kurtz, Ernest. *Not God: A History of Alcoholics Anonymous*. Center City, MN: Hazelden Publishing, 1979.

Langton, Kenneth P. "Peer Group and School and the Political Socialization Process." *The American Political Science Review* (1967): 751–58.

Ledgerwood, Alison and Shannon Callahan. "The Social Side of Abstraction: Psychological Distance Enhances Conformity to Group Norms." *Psychological Science* 23, no. 8 (2012): 907–13.

Ledgerwood, Alison, Yaacov Trope, and Nira Liberman. "Flexibility and Consistency in Evaluative Responding: The Function of Construal Level." *Advances in Experimental Social Psychology* 43 (2010): 257–95.

Lindquist, G. and S. Coleman. "Introduction: Against Belief?" *Social Analysis: The International Journal of Social and Cultural Practice* 52, no. 1 (2008): 1–18.

Liska, Allen E. "A Critical Examination of the Causal Structure of the Fishbein and Ajzen Attitude-Behavior Model." *Social Psychology Quarterly* 47, no. 1 (1984): 61–74.

Mele, Alfred R. "Akratic Feelings." *Philosophy and Phenomenological Research* 50, no. 2 (1989): 277–88.

Mercer, David. "Science, Legitimacy, and 'Folk Epistemology' in Medicine and Law: Parallels Between Legal Reforms to the Admissibility of Expert Evidence and Evidence-Based Medicine." *Social Epistemology* 22, no. 4 (2008): 405–23.

Mercier, Hugo. "The Social Origins of Folk Epistemology." *Review of Philosophy and Psychology* 1, no. 4 (2010): 499–514.

Merton, Robert K. *On Theoretical Sociology: Five Essays, Old and New*. New York, NY: Free Press, 1967.

Mortimer, Jeylan T. and Roberta G. Simmons. "Adult Socialization." *Annual Review of Sociology* 4 (1978): 421–54.

Murray, Thomas S., Vanessa L. Malcarne, and Kathy Goggin. "Alcohol-Related God/Higher Power Control Beliefs, Locus of Control, and Recovery Within The Alcoholics Anonymous Paradigm." *Alcoholism Treatment Quarterly* 21, no. 3 (2003): 23–39.

Netland, Harold A. "Professor Hick on Religious Pluralism." *Religious Studies* 22, no. 2 (1986): 249–61.

Nottelmann, Nikolaj. "The Analogy Argument for Doxastic Voluntarism." *Philosophical Studies* 131, no. 3 (2006): 559–82.

Oakley, I. T. "An Argument for Scepticism Concerning Justified Beliefs." *American Philosophical Quarterly* 13, no. 3 (1976): 221–28.

O'Hear, Anthony. "Belief and the Will." *Philosophy* 47, no. 180 (1972): 95–112.

Plato. *Protagoras*. Translated with notes by C. C. W. Taylor. Oxford: Clarendon Press, 1976.

Price, H. H. "The Inaugural Address: Belief and Will." *Proceedings of the Aristotelian Society* 28 (1954): 1–26.

Read, Jason. "A Genealogy of Homo-Economicus: Neoliberalism and the Production of Subjectivity." *Foucault Studies* 6 (2009): 25–36.

Roche, Timothy D. "The Perfect Happiness." *The Southern Journal of Philosophy* 27, no. S1 (1989): 103–25.

Roman, Paul M. "The Social Transformation of Alcohol Treatment." *Contemporary Sociology* 17, no. 4 (1988): 532–35.

Schneewind, Jerome B. "Natural law, Skepticism, and Methods of Ethics." *Journal of the History of Ideas* 52 (1991): 289–308.

Schneewind, Jerome B. "Voluntarism and the Foundations of Ethics." *Proceedings and Addresses of the American Philosophical Association* 70, no. 2 (1996): 25–41.

Simonton, Dean Keith. "Creative Thought as Blind-Variation and Selective-Retention: Combinatorial Models of Exceptional Creativity." *Physics of Life Reviews* 7, no. 2 (2010): 156–79.

Smith, Michael. "Minimalism, Truth-Aptitude and Belief." *Analysis* 54, no. 1 (1994): 21–26.

Surlin, Stuart H. and Thomas F. Gordon. "Selective Exposure and Retention of Political Advertising." *Journal of Advertising* 5, no. 1 (1976): 32–44.

Trompette, Pascale and Dominique Vinck. "Revisiting the Notion of Boundary Object." *Revue d'Anthropologie des Connaissances* 3, no. 1 (2009): 3–25.

Vähämaa, Miika. "A Group Epistemology is a Group Necessity: A Reply to Fallis and Mathiesen." *Social Epistemology* 27, no. 1 (2013): 26–31.

Vähämaa, Miika and Mark D. West. "'They Say One Thing and Mean Another'. How Differences in In-Group Understandings of Key Goals Shape Political Knowledge: An International Comparison of Politicians and Journalists." *Nordicom Review* 36 (2015): 19–34.

Weinstock, Stefan. "Victor and Invictus." *The Harvard Theological Review* 50, no. 3 (1957): 211–47.

Wheelis, Allen. *How People Change*. New York, NY: Harper and Row, 1973.

Winch, Peter. "The Expression of Belief." *Proceedings and Addresses of the American Philosophical Association* 70, no. 2 (1996): 7–23.

Wittgenstein, Ludwig. *Tractatus Logico-Philosophicus*. Rev. ed. London: Routledge and Kegan Paul, 1974.

Chapter 16

Empirical Social Epistemology

Addressing the Normativity of Social Forces

Miika Vähämaa

Social groups are undoubtedly fundamental to social life. As the core units of social life the many groups to which people belong function as the key media in the relation between knowledge and society at large. Social groups, as empirical units that organize thoughts and actions, are interesting places for research in social epistemology.

Traditionally, social groups have been the target of study in sociology and social psychology. These disciplines have focused on empirical social epistemological issues either through the brand of the sociology of science (e.g., Knorr-Cetina 1999; Pickering 1984; Rouse 2002) or through empirical social psychology of knowledge (e.g., Kruglanski et al. 2010; Moscovici 1981), research in personal epistemologies (e.g., King and Kitchener 1994, 2002; Perry 1970), and research in social cognition and persuasion (Nisbett and Wilson 1977; Pelham, Mirenberg and Jones 2002; Petty and Cacioppo 1986).

These traditions of research are characteristically far from philosophical social epistemology and even recent works seem to be conspicuously ignorant of the advances in social epistemology.

LACK OF OVERLAP BETWEEN PHILOSOPHICAL AND EMPIRICAL RESEARCH

In all fairness, it must be said that the philosophical schools have not done much to close the gap between empirical approaches taken in sociology and social psychology, and the purely conceptual analysis present their own epistemological work. What we are left with, consequently, are the philosophical theories of epistemology addressing knowledge as a social entity as if the domain of social was purely conceptual, or, at least better served without the troubles of empiricism.

Currently, there is not much overlap between philosophical epistemology, as portrayed either by the analytic school of social epistemology (Goldman 1999) and Audi (2010), or Fuller's (2002) normative social epistemology, and the empirical study of epistemology as a social praxis as lay epistemologies (e.g., Kruglanski et al. 2010), personal epistemologies (e.g., King and Kitchener 2002) and epistemic communities and goals (e.g., Fallis 2007; Knorr-Cetina 1999).

Both camps, the philosophical and the empirical, claim leading authority over epistemic social processes. Both approaches have important lessons for research in empirically oriented epistemology. In this chapter I first focus on the empirical insights and their normative relevance in epistemology. Then, toward the end of the chapter, I attempt to combine empirical ideas with a philosophical point of view.

EPISTEMIC RELEVANCE OF SOCIAL FORCES

The empirical school has systematically underlined the constraints of intentional thinking that people face at the social end of the epistemological spectrum. Empirical work has demonstrated that humans often operate with knowledge more intuitively than intentionally. Philosophical approaches, however, take it more or less as the premise that humans are intentional by definition and therefore there is no need to highlight the possible deficiencies of the actual—empirically founded—social processes and social dynamics. There is, however, important merit in the empirical findings.

Empirical research projects have cumulatively brought evidence of the relevance of social forces as epistemic norms. This means that the dynamics present in the group processes may at first appear as purely social means for social ends without anything particularly epistemological in those processes. Yet, a closer look reveals that the social dynamics of groups have the powers to importantly shape what is regarded as knowledge in groups and to define how things gain the status of knowledge.

To demonstrate the relevance and potential of the empirical findings, let us briefly consider the results gained in two prominent empirical fields: the lay epistemic theory of social psychology of knowledge and the automaticity theory in social cognition tradition.

LAY EPISTEMIC THEORY

Of the empirically orientated sociopsychological approaches of epistemology perhaps the most cited research program is the *lay epistemic theory* (LET), led by Arie Kruglanski (see Kruglanski et al. 2006, 2010). Kruglanski's (1989) research program for the past thirty years has been built under the umbrella of what he calls the LET. According to Kruglanski and colleagues, the research agenda of the LET consists of three different *specific* theories: the need for cognitive closure theory, the unimodel theory, and the epistemic authority theory (Kruglanski et al. 2010, 939–40). Most studied and cited is the theory of the need for cognitive closure. In the the authors' own words the theory posits the following (Kruglanski et al. 2010, 940):

> The need for nonspecific closure reflects a desire for a firm answer to a question, any answer, when compared to confusion and ambiguity. The need for specific closure reflects a motivation to reach a specific, personally desirable, answer to a question. Each need is assumed to vary in degree and lie on a continuum. Thus, one may desire closure strongly, mildly or not at all, even wanting to avoid it. Finally, both types of need for closure determine the length of the hypothesis generation and testing sequence.

The need for cognitive closure, Kruglanski and colleagues argue, is such a compelling aspect of individuals' need to make sense of the world that oftentimes any firm and acknowledged belief is hailed as knowledge in order to avoid continuous processing and reflection (Kruglanski et al. 2010, 940). Researchers have found that the need to get a cognitive closure or a need to avoid it is such a strong force in peoples lives that it can guide other social group processes. For instance, a person with a high need to get a cognitive closure—a need to just call something as steadfast knowledge—much favors strong authorities in groups and dislikes group debates that wax and wane without a clear closure.

Researchers hypothesize that the individual differences in need for cognitive closure may be one of the leading causes of societal unrest since some may desire a strong religious or political regime based on their cognitive needs while others, for the same reasons, are much opposed to any sort of authoritarian approaches (Kruglanski et al. 2006, 97). Kruglanski and colleagues (2006) conclude that one should not overlook the relevance of need for cognitive closure as an underlying mechanism influencing knowledge formation in groups.

AUTOMATICITY OF COGNITION

In the philosophical approaches of social epistemology, the connection between intentionality and language is commonly taken as the premise of epistemology. Yet, the research community in social cognition tradition is less convinced about the powers of human consciousness when it comes to forming knowledge in social settings.

For instance, Nisbett and Wilson (1977), Bargh, Chen, and Burrows (1996), and Pelham and colleagues (2002) present experimental cases that argue for the lack of human conscious reasoning in many important decision-making situations where knowledge is needed and formed. Experimental scenarios are typically quite casual, but nonetheless they pose an important question about the limitations of conscious reasoning in group contexts. If individuals tend to be driven by blind affect, then, how meaningful it would be to talk about an epistemic group process that requires conscious participation? Let us consider the experiment by Nisbett and Wilson (1977, 243–49) for an answer.

THE STOCKINGS EXPERIMENT

In Nisbett and Wilson's (1977, 243–49) experiment the test subjects had to choose from four identical pairs of stockings and evaluate which of them was of the highest quality. The research found a significant positional bias in the results. The test subjects were four times likely to choose stockings placed on the right in front of them. The reasons, however, that the test subjects gave for their choices were vivid examples of confabulation. No reference was made to the position of the chosen stockings. Rather, the test subjects described the qualitative differences in knot, sheerness, and so on. Upon querying about the possible positional bias the test subjects straightforwardly denied such possibility.

Nisbett and Wilson (1977, 243–49) hypothesized the results to indicate that when asked for explanation under social pressure in a group context, the subjects implicitly consulted cultural and personal a priori causal theories they had, which suited the situation nicely.

In other words, in the stocking scenario the test subjects tried to find reasonable explanations to their own behavior and in order to do so they drew rapidly upon their old mental schemas that resembled the present situation. The social goals of pleasing other people and to appear as rational in the eyes of the self and others were hypothesized as important social forces that made even the test subjects to believe that their observations and decisions were driven by conscious reason—not affective automaticity of cognition.

LATENT SOCIAL INFLUENCE AND KNOWLEDGE

The above findings may not at first glimpse seem to pose an important challenge to philosophical social epistemology. Clearly, one could argue, humans are not driven by blind automaticity in all social encounters or most of the time. Yet, it is important to consider the epistemic powers of quick mental processes. The reasons we give and the knowledge we form based on rapid or automated processes build on culture and social habits. At least every now and then all of us are influenced by our social desires toward appearing as reasonable and liked.

What I think is relevant here is the compelling view that reasoning and knowledge formation in groups can be influenced by latent social forces and, as such, these forces are theoretically interesting for social epistemology.

What if, the question emerges, the social forces—as those presented as the need for cognitive closure and automaticity of reasoning—are powerful enough to be regarded as a normative epistemic force? Should social epistemology regard social forces as a type of naturalized normativity that functions in group settings and enables people to better make sense of their social reality?

EMPIRICAL RESEARCH IN NORMATIVE LIGHT

If empirical epistemology has something to offer to a philosopher, it is the cumulative efforts made to show how social forces function normatively in social life. The trouble is that the epistemic norm-building nature of social forces is easy to regard as an epiphenomenon of social processes that do not weigh enough to catch epistemologists' full attention. At least part of the reason for this problem is the lack of bridging conceptual research at the interface of empirically drawn conclusions and conclusions taken from conceptual epistemological analysis.

Among the epistemologists the worry has been that empirical researchers do not fully address the issue of normativity—the norm building and maintaining quality—of their research methodology, premises, argumentation, and so on. However, if one is to pay any attention to the empirical research at all, the ontological and epistemological debate regarding the research approaches should be put aside at first.

Potentially a better starting point—in contrast to just fueling the philosophical versus empirical debate—for a critical scrutiny is the potential normativity that lies in the social processes per se. The reason for this is straightforward. If social forces really matter as much in the epistemic norm building as the empiricists paint them to, then it might be worthwhile reflecting whether social forces have some fundamental or axiomatic relevance in the pursuit of knowledge that is at once of empirical and philosophical relevance.

SOCIAL FORCES AND THE BEGINNING OF NORMATIVITY

Very much in the same way as every written argument must start with a letter, must every group start with an epistemology: at least some sort of lay theory of understandable and nonsense sentences, as suggested by Jürgen Habermas in communication theory and by Jan Smedslund in his work on psychologic (Habermas 1984; Smedslund 1988, 2012). In order for social gatherings to make sense to people there needs to be some sort of group epistemology right at the very genesis of a group. A group epistemology, thus, is a set of shared reference points in deciding what makes sense and what does not.

Agreeing on what makes sense is the beginning of normativity, or "proto-normativity," regarding knowledge in the social world. As the studies in automaticity of cognition suggested, social agreement can happen swiftly with just a latent reference to desirable norms (Nisbett and Wilson 1977).

To imagine how such an elementary level of epistemic praxis may evolve, we could think of two human beings meeting each other for the first time in order to achieve some shared meanings via communication. As the biological history has it, people have throughout time formed groups and sought out other people to meet individual needs through collective action.

A "CAVEMAN" EPISTEMOLOGY

The painted image of a caveman epistemology may be a silly example as such, but it does underline the connection between groups and group epistemologies. Epistemologically speaking, social groups function as the axiomatic basis for collective knowledge.

As the group evolves through social norms, it begins immediately to create further rules about the world "out there." Even in the eras before language in the contemporary sense, the atomic items of perception—birds, trees, food, and dwellings, as well as the presumable other people in the perceptible world—would have been given a set of meanings that stem not only from perceptions but from meanings within the social reciprocity the group members have.

People would not speak merely of "the big house," we would posit; they would speak of "the big house where the chief lives," a nomenclature with a social dimension. They would thus be likely to develop mutual trust and a felt sense of belonging together. The Ur-sprache would, in its most primitive form, have to be social, since it would have to be a joint effort; the Ur-sprache would be a group epistemology as well as a shared discourse.

PROTO-NORMATIVITY IN FOLKLORE

The caveman/woman scenario is actually a modified version of another well-known Western folkloric tale of the genesis of a group epistemology. The Christian Bible goes to some length to describe how early people encountered animals and plants and the like, and in the process of observing them, the First Man (אָדָם) gave them names in the sanctuary (גַּן עֵדֶ)—the first bits of social knowledge of the world. Adam, as the religious legend has it, formed a group epistemology with none other than God himself and communicated with God the names of various things that sounded good to him (or, as Genesis 1:27 has it, them; the use of אָדָם there is plural).

While there is no way, beyond various historical narratives, to know the actual historical development of early group epistemologies, the important point remains. Groups arise to form meanings, and in the process they craft rules for epistemic judgment.

NATURE OF NORMATIVITY IN SOCIAL INTERACTION

As studies in lay epistemologies (Kruglanski et al. 2006) as well as in automaticity of cognition (Nisbett and Wilson 1977) suggest, social norms emerge with ease in interaction. The nature of such interactive normativity is manyfold and draws upon culture, use of language, the individual needs for cognition and attempts to make sense of the self and others in a social setting.

Is such normativity, derived from social forces as they appear in culture, only valuable as a relativistic cultural process from a philosophical perspective? Could there be some theoretically compelling overarching principle connecting the various driving forces which bring and keep social groups together?

My answer goes as follows. The given review of the lay epistemological theory, the theory of automaticity of cognition, and a peek into how knowledge formation is represented in folklore are all conceptually connected. They portray social forces as goal oriented, that is, as teleological. Furthermore, they point out how social goals also help people to organize knowledge.

GOALS OF GROUPS

Groups' goals have much puzzled social science. First of all, the goal of making sense to others via language has been an important topic in social theory, perhaps most prominently in Habermas's (1984) work regarding the communicative action. Second, the goal of maintaining group coherence has been the target of research in empirical social psychology under the umbrella of social identity theory, promoted by Henri Tajfel and others (Tajfel 1982; Tajfel and Turner 1986). Third, research has posited that both language and group cohesion serve yet an another superior goal: the positive outcomes attached to purposeful interaction (Ryff 1989; Vähämaa 2013). This positivity has been a topic of both philosophical and empirical research (Annas 1993; Fredrickson 2009).

In a theoretical sense the main principle behind social forces is the pursuit of socially desirable and happy outcomes, as philosophical research in the morality of happiness (Annas 1993) and empirical positive psychology (Fredrickson 2009) assert. The exact content of such desires, according to Annas (1993, 2004) and Fredrickson (2009), is an empirical question. Yet, it is a question that is also a normative one. Important human goals of psychological well-being and philosophically understood well-being, or ευδαιμονία, go hand in hand (Ryff 1989). The main reason for this connection, once again, is language. Once people start to talk about happiness, they already take some rudimentary steps toward happy outcomes by relying on language and the prospect of the conversation to make sense (Habermas 1994, 111).

While pursuing the norms of well-being and happiness people necessarily rely on language (e.g., sense making) and group coherence (e.g., getting along with others). By fulfilling social goals in interaction, people also produce and circulate knowledge by talking and relying on common sense (Smedslund 2012) and by accepting beliefs due to group coherence (Kruglanski et al. 2006).

In what follows I provide a summary of the goal-oriented group processes from the vantage point of social epistemology.

SUMMARY OF GROUPS' EPISTEMIC POWERS

Thus far we have identified core group processes that contribute to knowing in a group context. These processes were conceptualized as teleological and having epistemic consequences. From the social epistemological view group processes that keep and bring groups together also allow circulation, acceptance, and validation of knowledge.

In summary, the following three social functions of groups are also the core properties that enable social groups to gain and circulate knowledge. Groups provide individually felt senses conducive to knowledge which are:

1. The sense that one possesses an ability to discuss and pursue truth, and to review and select different perceptions (following Habermas 1984, 11; 1994, 232; Smedslund 2012).

2. The sense of being a functional, accepted, and credible member of the group (following Kruglanski et al. 2006; Tajfel 1982, 21–22).
3. The sense of an ability to maintain personal affective stability and to achieve happiness (following Annas 1993, 43–46; 2004; Fredrickson 2009).

Together, all three senses act as social forces in group contexts and provide not only social functionality but represent also a pathway to organize knowledge in groups (see Vähämaa 2013). The conceptualization of social forces as *senses* instead of feelings or purely intellectual reasons highlights the dual nature of social forces. The presented features of interaction are all both intuitive and intentional and as such the normative basis of knowledge is rooted not only in language and interactive common sense but also in moral and philosophical pursuit of well-being and happiness.

The summary of group processes gives a reason to believe that while an individual in group context pursues some social end he or she might, as if accidentally, end up acquiring and circulating beliefs regarded as knowledge. On the other hand, the attempts to organize knowledge in group settings can be fully intentional. The individual's awareness of the social epistemological function varies, but the locus of action is the social group.

BIBLIOGRAPHY

Annas, Julia. *The Morality of Happiness*. Oxford: Oxford University Press, 1993.

Annas, Julia. "Happiness as achievement." *Daedalus* 133, no. 2 (2004): 44–51.

Audi, Robert. *Epistemology: A Contemporary Introduction to the Theory of Knowledge*. New York, NY: Routledge, 2010.

Bargh, John A., Mark Chen, and Lara Burrows. "Automaticity of Social Behavior: Direct Effects of Trait Construct and Stereotype Activation on Action." *Journal of Personality and Social Psychology* 71, no. 2 (1996): 230–44.

Fallis, Don. "Collective Epistemic Goals." *Social Epistemology* 21, no. 3 (2007): 267–80.

Fredrickson, Barbara. *Positivity*. New York, NY: Crown, 2009.

Fuller, Steve. *Social Epistemology*. 2nd ed. Bloomington, IN: Indiana University Press, 2002.

Goldman, Alvin I. *Knowledge in a Social World*. Oxford: Clarendon Press, 1999.

Habermas, Jürgen. *The Theory of Communicative Action: Reason and the Rationalization of Society*. Vol. 1. London: Heinemann, 1984.

Habermas, Jürgen. *The Past as Future*. Interviewed by Michael Haller, translated and edited by Max Pensky. Cambridge: Polity Press, 1994.

King, Patricia M. and Karen S. Kitchener. *Developing Reflective Judgment*. San Francisco: Jossey-Bass Publishers, 1994.

King, Patricia M. and Karen S. Kitchener. "The Reflective Judgment Model: Twenty Years of Research on Epistemic Cognition." In *Personal Epistemology: The Psychology of Beliefs about Knowledge and Knowing*, edited by Barbara K. Hofer and Paul R. Pintrich, 37–61. Mahway, NJ: Lawrence Erlbaum, 2002.

Knorr-Cetina, Karin. *Epistemic Cultures: How the Sciences Make Knowledge*. Cambridge, MA: Harvard University Press, 1999.

Kruglanski, Arie. W. *Lay Epistemics and Human Knowledge: Cognitive and Motivational Bases*. New York: Plenum, 1989.

Kruglanski, Arie W., Edward Orehek, Mark Dechesne, and Antonio Pierro. "Lay Epistemic Theory: The Motivational, Cognitive, and Social Aspects of Knowledge Formation." *Social and Personality Psychology Compass* 4, no. 10 (2010): 939–50.

Kruglanski, Arie W., Antonio Pierro, Lucia Mannetti, and Eraldo De Grada. "Groups as Epistemic Providers: Need for Closure and the Unfolding of Group-Centrism." *Psychological Review* 113, no. 1 (2006): 84–100.

Moscovici, Serge. "On Social Representations." In *Social Cognition: Perspectives on Everyday Understanding*, edited by Joseph P. Forgas, 181–209. London: Academic Press, 1981.

Nisbett, Richard E. and Timothy D. Wilson. "Telling More Than We Can Know: Verbal Reports on Mental Processes." *Psychological Review* 84, no. 3 (1977): 231–59.

Pelham, Brett W., Matthew C. Mirenberg, and John T. Jones. "Why Susie Sells Seashells by the Seashore: Implicit Egotism and Major Life Decisions." *Journal of Personality and Social Psychology* 82, no. 4 (2002): 469–87.

Perry, William G. *Forms of Intellectual and Ethical Development in the College Years—A Scheme*. New York: Holt, Rinehart and Winston, 1970.

Petty, Richard E. and John T. Cacioppo. *Communication and Persuasion: Central and Peripheral Routes to Attitude Change*. New York: Springer, 1986.

Pickering, Andrew. *Constructing Quarks: A Sociological History of Particle Physics*. Chicago: University of Chicago Press, 1984.

Rouse, Joseph. *How Scientific Practices Matter: Reclaiming Philosophical Naturalism*. Chicago: University of Chicago Press, 2002.

Ryff, Carol D. "Happiness is everything, or is it? Explorations on the meaning of psychological well-being." *Journal of Personality and Social Psychology* 57, no. 6 (1989): 1069–81.

Smedslund, Jan. *Psycho-Logic*. New York: Springer-Verlag, 1988.

Smedslund, Jan. "Psycho-Logic: Some Thoughts and After-Thoughts." *Scandinavian Journal of Psychology* 53, no. 4 (2012): 295–302.

Tajfel, Henri. "Social Psychology of Intergroup Relations." In *Annual Review of Psychology*, edited by Mark R. Rosenzweig and Lyman W. Porter, 1–39. Palo Alto, CA: Annual Reviews, 1982.

Tajfel, Henri and John C. Turner. "The Social Identity Theory of Intergroup Behaviour." In *Psychology of Intergroup Relations*, edited by Stephen Worchel and William G. Austin, 7–24. Chicago, IL: Nelson-Hall, 1986.

Vähämaa, Miika. "Groups as Epistemic Communities: Social Forces and Affect as Antecedents to Knowledge." *Social Epistemology* 27, no. 1 (2013): 3–20.

Chapter 17

On Feminist Epistemology

The Fallibility of Gendered Science

Diana Rishani

The claim that science is free from biases and epistemological contaminations is a claim that must be examined thoroughly, especially in terms of context. The contextualization aims to deconstruct the practice of science with respect to notions of power relations, capitalism, and the default/Other dichotomy among others. Feminist epistemology focuses not just on the place of "women" in science, but also on all the Other unprivileged groups of society, and it does so by analyzing the social, political, and ideological context science is made in. This analysis critiques the institution of science as a whole and therefore brings forth questions on the nature of objectivity and whether it is achievable or not. This brief chapter will discuss how science's claim of objectivity is deemed to be fallible in the light of feminist epistemology, and will instead present a notion of objectivity in the light of context and intersectionality.

The Other were rarely the agents of science nor even the subjects of it, instead they were the invisible consumers of whatever it produces. Objectivism erased all social values from the process of science and instead imagined itself building a neutral realm in which empirical data is waiting to be snatched. By making this value-free claim, science is therefore asserting the already value-loaded framework it operates in as a doubtless default. The white, male, able-bodied, and western-centric science establishes a hegemony over knowledge production and in turn a hegemony over body politics, economics, and international relations.

> The conception of value-free, impartial, dispassionate research is supposed to direct the identification of all social values and their elimination from the results of research, yet it has been operationalized to identify and eliminate only those

> social values and interests that differ among the researchers and critics who are regarded by the scientific community as competent to make such judgments. (Harding 2008, 744)

The historical context of colonialism allowed a power structure to emerge, which is perpetuated by capitalism. That structure was/is an extension of white, male, western ideologies imposed on the world as the "truth" and science, in turn, works within these boundaries.

However, if science is then deemed as a hegemonic power of the androcentric west, and its objectivity is then refuted, does that entail an entire rejection of the notion of objectivity as it relates to science? Embracing a skeptic view and denying the existence of a uniform external world enable one to easily accept relativism, which itself holds many dangers. Relativism allows scientists to admit their faulty and centric ways and just carry on with it. It gives the option of opting out of discussions where worlds, struggles, and oppressions are different. Cultural relativists dangerously tread on the boundaries of separation between the respect for human rights and respect for cultural diversity. Therefore, awareness of the different context is not enough due to the fact that the western androcentric science operates in the world and affects it respectively—leaving the Other to operate one's own science is not a valid alternative since the production of science already exists in an institutionalized power structure. Instead, one should accept the limitations of awareness, as the fish does not know what water is, and aim toward a more holistic, inclusive, intersectional science based on a probability conscious of the context and power structure it exists in. This transforms the concept of objectivity from a singularity to a spectrum: a claim can be "more objective" and a more real reflection of reality than another.

Feminist epistemology operates within this spectrum and thus questions the essentialist claims made by the west. Anthropology began as a colonial science aimed at studying "the native," it facilitated the perpetuation of the already existing racist claims of the west, and in turn, science has occupied that same position. Feminist epistemology tackles the essentialist claims that readily identify "women" and continues to reinforce the constructed image of one. Science produces value-loaded "knowledge" into a capitalist system, framed in terms of patriarchy. The business of "woman making" is a highly profitable one in which "women" (or people who identify themselves as such), both privileged and the Othered, feed the system financially and ideologically. Marxist feminism is a form of epistemology that acknowledges the power structure set up by capitalism and patriarchy and aims to deconstruct the ideological manifestations in terms of that structure.

Feminist standpoint theory, among others, is a form of feminist epistemology.

> Feminist standpoint theorists make three principal claims: (1) Knowledge is socially situated. (2) Marginalized groups are socially situated in ways that make it more possible for them to be aware of things and ask questions than it is for the nonmarginalized. (3) Research, particularly that focused on power relations, should begin with the lives of the marginalized. (Bowell 2014)

The theory situates science in the structure it was produced in to allow critical analysis of its nature, and in turn understand its effect on the Other's life. Feminist standpoint theory begins with a descriptive project and ends with a normative one. Its intersectional approach appeals to the pragmatic nature of science. Capitalism has always used science to establish new means of exploitation in order to increase profit and continues to do so. The monopoly of science by the hands of the capitalists has allowed the production of new sciences for the aims of profit. Marxists continue to claim that the machine is not innately exploitative; it is rather the capitalist's application of it that is exploitative. Instead of relieving the workers from intense physical labor, it just establishes a more monotonous and strenuous type of work. As long as science continues to operate in a power structure that demands new tools, science can never claim to be an "objective" state.

However, since feminist standpoint theory already exists in an androcentric western structure, it could be critiqued by saying it is a close resemblance to an Othering project, one with a relativist sense. Feminist standpoint theory and its projects "conflict because the notions involved are perfectly coherent with the maintenance of elitist knowledge production and systems" (Harding 2008, 751). Acknowledging different agendas and claiming to truly understand the Other only to then prefer the already established elitist project is just another form of hegemony. It aims to integrate the Other into the power structure by pretending to address the oppressions and differences found, rather than completely deconstructing the structure. The Other is given a foot note, an imaginary agency that complies to the hegemonic and preexisting scientific institution. The Other is still the subject of study, the deviation from the hegemonic norm. The production of knowledge carries on with the assumption of objectivity, and the default nature of the elite is never subject to explanation and questioning.

Nonetheless, the feminist standpoint theory project should not be put at a halt. Although the risk of Othering may be present, especially at the hands of those who hold the positions of power in the hegemony, this intersectional form of knowledge production should not be abandoned. Due to its duality of description and normativity, the feminist standpoint theory appeals to the necessity of changing the androcentric western structure. The theory understands how science, as a pragmatic practice, can be molded to benefit those it has long wronged. The theory situates the agents of knowledge production in their respective context and allows a meta-analysis of this knowledge; it does not treat it as an unquestionable absolute truth. Ideally it allows agency to be granted to the Other in order to produce knowledge, supposedly set to be as equal to that of the hegemony. However, academia is structured in a way that does not allow the Other to enter without institutionalization, without the removal of the Other from the "native" self in order to comply to the structure. Feminist standpoint theory acknowledges such claims by contextualizing the given, and in turn sets to question the state of "objectivity." The theory allows science to move along the spectrum of "objectivity" toward a stronger form of it—the theory is a support of constant questioning. Even though at instances in academia it can be claimed by the hegemony, the theory should not be given up on for the importance of its intersectional approach.

Science can never be a "pure and objective" form of knowledge. If it was, then its value would be questionable due to having its pragmatic nature compromised. If it cannot be related to a real context, then the experience generating it and its application would be futile. "To insist that no judgments at all of cognitive adequacy can legitimately be made amounts to the same thing as to insist that knowledge can be produced only from 'no place at all'" (Harding 2008, 750). Feminist standpoint theory aims to further push knowledge production along the spectrum of objectivity via its intersectional approach. It cultivates the agency of the Other and grants an opportunity to question the hegemonic process of knowledge production that so deeply influences everyone's life.

BIBLIOGRAPHY

Bowell, Tracy. "Feminist Standpoint Theory." *Internet Encyclopedia of Philosophy.* ISSN 2161-0002. www.iep.utm.edu/fem-stan/, November 15, 2014.

Harding, Sandra. "Strong Objectivity and Socially Situated Knowledge." In *The Feminist Philosophy Reader*, edited by Alison Bailey and Chris J. Cuomo, 741–56. Boston, MA: McGraw-Hill, 2008.

Chapter 18

The Cost of Being Known

Economics, Science Communication, and Epistemic Justice

Fabien Medvecky

Immortality is possible; we know this because it exists. *Turritopsis dohrnii* is a jellyfish capable of reversing from a fully sexually mature medusa stage back to a polyp stage before maturing again from polyp to medusa (Miglietta and Lessios 2009). By doing so repeatedly, it can seemingly cheat biological death (but it can still get eaten and such like)! So how can I get to know more about this, and how did I get to find out about *Turritopsis dohrnii* in the first place? I have invested time into knowing about biology, I have invested resource on books, computers, and so forth to get access to these facts, and if I want to know more, I will have to invest more.

Knowledge comes at a cost, a cost in the classic economic sense, a cost in terms of resources such as time and money. If I want to know *X* then I will have to invest resources into the act of getting to know *X*. We can think of knowledge as costing the "knower" something; costing those seeking the knowledge. But there is another way in which we may think of the cost of knowledge, and this is the cost to those wanting *X* to be known, the cost to the distributors of knowledge. Distributing knowledge takes resources, so how does this cost of distributing knowledge affect what we know and what perils we might face in acquiring new knowledge as a result of this cost? In this chapter, I want to concentrate on scientific knowledge and the cost of its communication and dissemination. I will argue that the variability and discrepancy in economic resources available to scientific endeavors and organizations leads to a biased view of what the current state of scientific knowledge is; a bias we should think of as a form of epistemic injustice in the vein of Miranda Fricker's definition of the term. This raises deep issues given that funding for science, scientific endeavors, scientific organizations, and science communication is deeply rooted in our social structures. I will briefly consider two specific cases that comes to bear on epistemic justice: private funding for science and the prominence of nature documentaries.

THE ECONOMICS OF KNOWLEDGE

Science is clearly related to knowledge (Fuller 1987). What exactly is this relation and how we should define it is not settled, but for my purpose, I will take science, and particularly scientific research, to be a knowledge-producing activity (among other things). Economic forces play an important role in shaping the path of this knowledge production (Stephan 2012). But economic forces also play a key role in the way this produced knowledge is acquired and distributed and in determining who gets access to it. This is because knowledge, or at least its acquisition and distribution comes at a cost, an economic cost.

When I say knowledge comes at an economic cost, I do mean monetary cost, such as the cost of buying books, laptops, or university courses. But I also mean other nonmonetary economic costs, such as time and effort. Some of these costs are obvious and direct—how much I paid for the latest Piketty book, *Capital in the Twenty-first Century* (2014). Other costs are opportunity costs; what I had to give up or didn't get to have so as to enjoy what I did get—the money I spent on the latest Piketty book was money I was not able to spend on my new subscription to *Nature* and the time I spent reading Piketty's book was time I was not able to spend reading more articles on social epistemology.

As a potential knower, these are costs we take seriously. We decide which university course we undertake—both in term of content and in term of institution—based not only on our interests in the topic or preference for the institution, but also because of costs and potential returns (Clotfelter 1996). Likewise, due to the scarcity of time and of funds, we make decisions over what we read, watch, and listen to in order to gain further knowledge, be it about philosophy or about latest findings in ecology. Put simply, if I want to gain some new knowledge, I will have to invest some of my resources into the act of acquiring that new knowledge. As potential knowers, costs are important in determining what knowledge we consume. Indeed, it may be useful to think of ourselves (in our capacity of potential knowers) as knowledge consumers, who, like all consumers, are sensitive to costs, monetary or otherwise.

But if we are to think of potential knowers as knowledge consumers, then the other side of this "knowledge market" are the knowledge suppliers or knowledge distributors (Boulding 1966). By knowledge distribution, I mean those activities meant to distribute information to knowledge consumers, and which, if the activity comes to fruition, will lead to the acquisition of new knowledge for the knowledge consumers. Importantly, I am solely interested in clearly intentional knowledge distribution. Since my interest is with knowledge produced out of scientific activities, our knowledge distributors of concern are what is termed science communicators, and the activities of concerns are those activities considered science communication activities.

As with knowledge consumers, knowledge distributors face economic costs. Distributing knowledge takes resources, both monetary and otherwise. These resources or costs can be both direct costs—how much funding is required to put the museum exhibition of interest together—or they can be indirect costs such as opportunity costs—the display that had to be given up so as to have sufficient funds and resources for the exhibition to go ahead. Knowledge distributors face scarce resources; there is neither sufficient time nor sufficient funds to distribute all of the knowledge produced by all scientific activities, so like knowledge consumers, knowledge distributors make decisions over how much they invest in each respective knowledge distribution activity. Some knowledge distribution activities are allocated more resources and, hopefully, have a wider reach, while other activities receive fewer resources and, all things being equal, have a lesser reach. But it is not just knowledge distributors who make the decisions as to how much is invested in each science communication activity; there are more substantial socioeconomic forces at play here because not all science communicators begin with an equal or commensurable set of resources for distributing knowledge, and this leads to some form of epistemic injustice.

THE ECONOMICS OF SCIENCE

Few would doubt the importance of economic forces in shaping the course of scientific research. Economic forces play a large part in the production of research from forcing collaboration on large, expensive projects that no single participant can afford on their own, such as the large hadron collider, to channeling or "guiding" research interest by the lure of potential financial returns such as mining or pharmaceutical research. There are rewards for doing science, such as salaries, grants, intellectual property, start-ups and consulting, and economic forces also play a role in forging collaborations; well-funded centers have large cohorts of post-docs, less well-funded centers use more research students. Indeed, in her book *How Economics Shapes Science* (2012), Paula Stephan discusses incentives and efficiency in scientific knowledge production. She suggests we need to look at our processes (especially, our incentives structures) and restructure them so as to allocate resources more efficiently. But efficiency isn't the whole story or the only possible take on it. We might also want to think about allocating them more equitably or more justly.

Resources are unevenly distributed across the sciences, both across endeavors and across organizations. Some countries are better resourced than others; some centers or institutions are better funded than others; some researchers are better paid than others; some projects or research programs receive more grants than others; and so forth. For example, the European Space Agency's (ESA) budget for 2009 was US$4.870 billion (ESA 2009). This is similar to the budget allocated in the United States to the National Cancer Institute (NCI) that year (US$4.967 billion; NCI 2009 Fact Book). By comparison, the US space agency, NASA, had a budget of US$17.770 billion for 2009 or just over three and half times more than either the ESA or the NCI (NASA 2011). Inequality among scientific organizations is nothing new, but it is something important for us. Variability in funding and resources is not only a determining factor of the kind of research that is being carried out—NASA can carry out more expensive research than the ESA or the NCI—it is also a factor in determining how the knowledge produced by each of these organizations' research will be distributed, communicated, and made public.

A well-funded organization has the capacity to distribute its knowledge more broadly than a less well-funded one, and, as a result, gain greater social recognition. For example, a recent survey in Australia found that the best-known Australian scientific achievement was the cochlear implant, and by a long margin—17 percent for cochlear, with the next closest, the cervical cancer vaccine, on a measly 8 percent (Searle 2014). Comparing the funding capacities and the current state of these two research programs is suggestive of why one is so much better known than the other.

The continuation of the cochlear implant program started by Dr. Graeme Clark (now carried on by Cochlear Limited) has an annual revenue of A$752 million (Cochlear Limited 2013). The continuation of the cervical cancer program started by Professor Ian Frazer (carrying on at the Translational Research Institute at the University of Queensland) has a total (not annual) budget of $354 million, less than half of Cochlear's annual revenue (The Atlantic Philanthropy 2009). Clearly, neither of these is poorly funded, but there is still a clear discrepancy, both in terms of resources and in terms of knowledge distribution. The concern is that the discrepancy in knowledge distribution leads to a bias in our social view of the current state of scientific knowledge, and this bias risk is being unjust, epistemically unjust.

FUNDING EPISTEMIC INJUSTICE

There are two quite different concepts we may be referring to when discussing epistemic injustice (Coady 2010). The first is in terms of distributive justice; whether all parties in the process of knowledge distribution are treated fairly (under some definition of fair) with regard to access or allocation of knowledge, or information. This is not the sense of epistemic justice I am interested in here. The second concept of epistemic injustice, the one I am interested in, is with regard to testimonies and credibility. This form of injustice stems from treating some entity unfairly as a knower, as a source of information or knowledge; what Fricker (2003, 164) describes as "prejudicial dysfunction in testimonial practice." Injustice in this sense comes from the hearer's action and the inappropriate degree of credibility the hearer gives to the speaker as a source of knowledge. As an example, Fricker mentions a scene in *The Talented Mr Ripley*. Recall that in the story, Tom Ripley had befriended Dickie Greenleaf before killing him and covering up the murder by making it seem like a suicide. In one of the final scenes, Herbert Greenleaf (Dickie Greenleaf's father) dismisses Dickie's fiancée (Marge Sherwood)'s claims that Dickie was in no way suicidal—in fact, she has the correct suspicion that Tom Ripley is responsible for his death. Marge's testimony is prejudiced on the basis that she is a hysterical woman in mourning for her sweetheart, and was dismissed on those grounds. This is an injustice because it "has the effect of undermining Marge as a possessor of knowledge about the lover she had been living with for some time" (168). Epistemic injustice is a prejudicial take on sources of knowledge's credibility, not because of the quality of knowledge those sources possess, but because of some other basis for prejudice. The inequality in resource distribution across various scientific organization and scientific endeavors, and the effect this has on their capacity to share and communicate their information, is one mechanism by which such prejudice can come to pass.

Much, if not all, of the discourse on epistemic injustice has focused on injustices ensuing from negative prejudice against the speaker. Marge Sherwood is treated unjustly as a possessor of knowledge because her knowledge claims are dismissed as a result of negative prejudices about "women in her situation." But prejudice is not always negative—it is not only unjustly dismissive or undermining of some subject's credibility. Prejudice can also be positive prejudice—it can be unjustly acceptive or supportive of some subject's credibility. As Fricker explains, the "prejudice results in the speaker's receiving more credibility than she rationally deserves—credibility excess—or it results in her receiving less credibility than she rationally deserves—credibility deficit" (164). Indeed, Herbert Greenleaf not only dismisses Marge's claims about Dickie's murder, but also actively endorses Tom Ripley's version of the events on the grounds that Ripley is a more credible possessor of knowledge; in his words, "what a man may say to his sweetheart and what he'll admit to another fellow" (Minghella 2000). With regard to the variation in resources available to scientific organizations and scientific endeavors, it is this positive prejudice, this credibility excess, which I think we ought to be concerned about.

Scientific organizations that are substantially better resourced than most can give themselves a better public profile. They can be better known, better recognized, and better socially established. With this recognition, the better-resourced organizations' messages are also better heard. They might be heard at the expense of other organizations or endeavors, other organizations or endeavors that are just as socially important, or possibly more so. They are also likely to benefit from greater credibility as a result of their increased prominence in the social sphere; they benefit from positive prejudice. This is especially important when there are conflicting or competing claims or projects to be considered, or when further resource allocations are at play. For example, consider debates in conservation management, such as the controversy over ecological triage, where some university-based research centers hold conflicting views to the United Nations environment program (Arup 2010; Bottrill et al. 2008). Or consider research in the mining and energy sector, where conflicting social and scientific forces often clash. In such contexts, the various participants' capacity to be heard, to be granted credibility, is inequitable because some party benefits from credibility excess as a result of their superior funding and resources, leading to epistemic injustice.

These injustices stem from complex social trends and structures that determine the resources and funding available to various scientific organizations and endeavors, and their capacity to distribute that knowledge through society, to communicate their projects, their outcomes, and so forth. And as Medina (2013) argues, "We have to pay attention to the social trends that may affect in direct or indirect ways what happens in the particular interaction and how participants perceive and appraise each other" (59). The social structures and pathways that lead to inequitable resource allocation, which generates epistemic injustice, have not been discussed yet. I have simply assumed that resources are "allocated" across various scientific organizations and to their respective science communicators, but this is clearly oversimplified. While much of scientific research is funded by governments and to that extent the allocative assumption just mentioned might resonate, funding sources for science and science communication is more complex than this (Broad 2014). One recent trend particularly relevant to the discourse on epistemic justice is the increasing share of funding coming in from philanthropists.

While governments still account for the largest share of scientific funding in the western world, an increasing amount of resources is now coming in from the private sector. This is causing some concern, concern largely over epistemic justice. Governments have, to a greater or lesser extent, let the scientific community determine what research is being carried out. However, philanthropist funding much more actively shapes the direction of research. The fear is that "as they are targeted only towards particular projects, they have the potential to skew the distribution of research projects being proposed and executed, by pulling resources from other, less 'popular' areas" (Nature Neuroscience 2008). And this skewing affects all aspects of the projects, including their communication. It affects their communication because well-funded projects are better resourced to distribute their knowledge and benefit from greater social recognition, giving them credibility excess.

On the other economic hand, not all science communication and knowledge distribution activities need to rely on external funding; some can generate income. Documentaries are a paradigm case where the distribution of knowledge is affected by the prospect or possibility of financial returns. *March of the Penguins*, for example, made $127,392,693 at the box office (Box Office Mojo 2014). Yet, not all aspects of science make for popular or successful documentaries. Indeed, documentaries are very heavily skewed toward the biological sciences (mostly nature documentaries), with a light sprinkle of physics. Of the top twenty most popular science and nature documentaries, seventeen were nature documentaries and the remaining three were about physics (IMDB 2014). Not one about chemistry or mathematics. Putting aside complex questions about the sociality of what makes a popular or successful documentary or why we should strive for such outcomes, the bias in the types of documentaries that do generate income also act as a way of skewing the social construction of science, with biological fields looming large. Indeed, it should come as no surprise that in the United States, there are 2.58 times more graduates with majors in the biological science than in the physical science (Siebens and Ryan 2009). The concern here is that biology (and related environmental sciences) benefits from credibility excess, making its research, endeavors, and so forth appear so much more important than that of other scientific fields, and possibly more than it "rationally deserves," to use Fricker's (2003) phrase.

CONCLUSION

Scientific research is a knowledge-producing activity, and it is also a social activity. It has its own sociality and it operates within the larger social sphere from which it gains funding and with which it shares its knowledge. To share the knowledge produced by scientific activity, scientists and scientific organizations communicate or distribute this knowledge to publics of potential knowers. The act of distributing knowledge comes at a cost, and the complexities of who pays for this cost, how much should be spent on it, and who should benefit from it raise substantial questions of justice, especially epistemic justice. I have made the point that inequitable resource allocation means that some scientific organizations are better resourced and benefit from increased social recognition and the ensuing credibility excess, a form of epistemic injustice. There are a number of ways in which this credibility excess may be exacerbated—the source of the funding, whether from government or from private philanthropy; the potential financial returns from the act of distributing knowledge; and so forth. While economic forces will always affect knowledge distribution, we socially construct these economic forces, and we should do so with a full awareness of the implications of our economic constructs on epistemic justice.

BIBLIOGRAPHY

Arup, Tom. "Too Early for Nature 'Triage': Scientist." *The Age*, July 1, 2010.

Bottrill, Madeleine C., Liana N. Joseph, Josie Carwardine, Michael Bode, Carly Cook, Edward T. Game, Hedley Grantham, Salit Kark, Simon Linke, Eve McDonald-Madden, Robert L. Pressey, Susan Walker, Kerrie A. Wilson, and Hugh P. Possingham. "Is Conservation Triage Just Smart Decision Making?" *Trends in Ecology and Evolution* 23 (2008): 649–54.

Boulding, Kenneth. "The Economics of Knowledge and the Knowledge of Economics." *American Economic Review* 56, no. 1/2 (1966): 1–13.

Box Office Mojo. *March of the Penguins* (2014). www.boxofficemojo.com/movies/?id=marchofthepenguins.htm.

Broad, William. "Billionaires With Big Ideas Are Privatizing American Science." *New York Times*, March 15, 2014.

Clotfelter, Charles. *Buying the Best: Cost Escalation in Elite Higher Education.* Princeton, NJ: Princeton University Press, 1996.

Coady, David. "Two Concepts of Epistemic Injustice." *Episteme* 7, no. 2 (2010): 101–13.

Cochlear Limited. "Cochlear Full Year Financial Results for Year Ended 30 June 2013 in Line with Guidance." *ASX and Media Release*, Sydney 2013.

European Space Agency. "ESA Budget for 2009." esamultimedia.esa.int/docs/corporate/ESA_2009_Budgetsweb.pdf.

Fricker, Miranda. "Epistemic Justice and a Role for Virtue in the Politics of Knowing." *Metaphilosophy* 34, no. 1/2 (2003): 154–73.

Fuller, Steve. "On Regulating What is Known: A Way to Social Epistemology." *Synthese* 73, no. 1 (1987), 145–83.

IMDB. "The Best Science and Nature Documentaries" (2013). www.imdb.com/list/ls057139208/.

Medina, José. *The Epistemology of Resistance.* New York: Oxford University Press, 2013.

Miglietta, Maria and Harilaos Lessios. "A Silent Invasion." *Biological Invasions* 11, no. 4 (2009): 825–34.

Minghella, Anthony. *The Talented Mr Ripley: A Screenplay.* London: Methuen, 2000.

NASA. 2011 Budget Overview (2011). www.nasa.gov/pdf/420990main_FY_201_%20Budget_Overview_1_Feb_2010.pdf.

Nature Neuroscience. "No Science Left Behind." Editorial 11, no. 10, October 2008.

Piketty, Thomas. *Capital in the Twenty-first Century*. Cambridge, MA: Harvard University Press, 2014.

Searle, Suzette. *How Do Australians Engage with Science? Preliminary Results From a National Survey*. Australian National Centre for the Public Awareness of Science (CPAS), The Australian National University, 2014.

Siebens, Julie and Camille Ryan. "Field of Bachelor's Degree in the United States: 2009." *U.S. Census Bureau*. U.S. Department of Commerce, Economics and Statistics Administration, Maryland, 2012. www.census.gov/prod/2012pubs/acs-18.pdf.

Stephan, Paula. *How Economics Shapes Science*: Cambridge, MA: Harvard University Press, 2012.

The Atlantic Philanthropies. "Translational Research Institute Queensland Receives Record $50M gift" (2009). www.atlanticphilanthropies.org/news/translational-research-institute-queensland-receives-record-50m-gift.

U.S. Department of Health and Human Services. *National Cancer Institute 2009 Fact Book*. Bethesda, MD: Office of Budget and Finance, National Cancer Institute, National Institutes of Health, 2009.

Chapter 19

Social Epistemology, Dialectics, and Horizontal Normativity

An Introduction to the Theory of Natural Authority

Pedro Saez Williams

If my intervention in the present volume could be summarized in a single statement, this would read, "The future of social epistemology lies in furthering its commitments to naturalism." This is, I claim that the fortune of all social epistemology—especially the epistemologically naturalist version developed by Steve Fuller—lies in its commitments to naturalism. In this same vein, I hold that its misfortune lies there, where it allows itself to depart from these commitments.

Fuller's "naturalist social epistemology" can be understood as an ambitious philosophical effort to produce a normative account for the conduct of systematized knowledge production, from a naturalist understanding of knowledge. A difficult and ambitious endeavor, if one considers that normativity has traditionally assumed some sort of metaphysical firmness for its guiding principles, and a naturalist rendition of "knowledge," "epistemic validity" or any other posited phenomenon of "epistemic" relevance, reduces these to overall metaphysical contingency.

The latter, furthermore, appears to be a general consequence of the "naturalization" of phenomena that were previously explained by alter-scientific means. This is, in order to avoid circular explanations (i.e., *x* is *x*, on account of *x*) "science" (or natural philosophy) accounts for the ontological nature of the *explanandum* in virtue of posited relations to something else. That which falls under the purview of science ends up being understood as caused, or otherwise ontologically determined by other phenomena and entities (contextual contingency).

In the case of "knowledge," metaphysical contingency is continuous with an understanding of general epistemic revisability and changeability. This is, the notion that all knowledge—including what was previously considered as the a priori principles that guide, prescribe, or provide the basis for knowledge production and/or epistemic judgment—is contingent to the context of its genesis and/or reception, implies that it does not hold a firmer metaphysical (epistemic) grounding than the rest of "nature." It is therefore, also, transient or subject to change, either on account of scrutiny or otherwise (see Quine 2002, 1951; Fuller 1992).

Epistemology (including philosophy of science) "has traditionally inhibited second-order impulses to find the nature of inquiry" on account of a "need [to identify] stable first-order enquires" (Fuller 2002, xiv). In other words, efforts at an epistemology have traditionally severed that which concerns science's normative dimension (i.e., the principles for that guide practice, epistemic judgment, and organization) from that which concerns its practice; the former being cast as "basic" or "pure," occupying a higher place in the epistemic hierarchy that locates them out of bounds of revision of empirical practice (vix–xv).

When a set of principles (or a "form of knowledge") that is posited as "basic," "pure," "firmer," "universal," "neutral," or otherwise situationally (contextually) uncontingent seem to justify and/or provide legitimacy to a social hierarchy, or to any social arrangement in which a specific agent (a) is able to limit and/or influence the capacity of action of another (b), without the latter (b) being able to scrutinize this process of decision making, it is common for naturalism to denunciate it as serving the role of an "ideology." This is, of serving as a justificatory explanation "designed to prevent criticism and revision" (Fuller 2002, 34). Notable examples of transcendentalist arguments accused of fulfilling this role are the Marxist (1970) denunciation of liberalism (and/or self-evident universal individual liberties) and the Lockean ([1689] 1988) challenge upon the Divine Right of Kings.

For Fuller (i.e., 2000, 2002, 2007; Fuller and Collier 2004) epistemic/semantic/cognitive transcendentalism operates precisely in such a way. It justifies, legitimates, and deters the scrutiny for the construct of "scientific experts," that is, individuals whose exercise of "cognitive authority" (i.e., "expertise") may affect third parties, and whose "validity" is independent of the participation of the latter. Taking seriously the empirical challenges to that which sustains such a notion (i.e., Barnes 1977; Latour and Woolgar 1986), leads to the consideration that the observed cognitive hierarchy of the scientific establishment (i.e., academic "ranks" and boundaries between disciplines) may be organized in such a way that allows the upper echelons of said hierarchy to act, under assumptions of legitimacy, as "judge and jury" in regard to practical, economical, and political matters that concern society at large (see Fuller 2000, 2001, 2002; Fuller and Collier 2004).

Identifying this situation as a pressing political—but most importantly socio-epistemological—problem, Fuller's program seeks to address the situation by prescribing a new set of normative principles for the organization of science. The project basically involves reorganizing all scientific efforts upon the principles of "the civic republican tradition of democracy." In practice, this would translate into the elaboration of a "constitution of science" whose prescriptions would provide the necessary social and political scrutiny for cognitive authority that seems to be unnecessary for traditional epistemology and absent in the current status quo. Such a regime would hypothetically force the scientific hierarchy to compete against each other in ways that they could previously dodge, and to subject their claims, and findings to members of the scientific community as a whole, as well as to those that might have a direct interest in the production of knowledge (i.e., members of the lay public). The hypothesis being that such a regime would transform the normative dimension of science from its current "geometrical" model, in which the normative component is understood as located in a higher (metaphysical/epistemic) ground, into a "dialectical" model in which a metanormative form (i.e., epistemic republican constitutionalism) allows the normative dimension (i.e., principles for production and epistemic judgment) to be in constant conflict with the empirical dimension or the practice of science: Locating, therefore, all knowledge within the possibility of scrutiny and change, and subjecting, thereby, all cognitive hierarchy to the same conditions. In short, leading to the "perpetual project of preventing any form of knowledge from becoming a vehicle of power" (see Fuller 2000, 2002; Fuller and Collier 2004).

The question is whether Fuller's republic of science would be able to achieve such a state of affairs. It would initially seem so.

In practice, constitutions that purport to follow principles such as those endorsed by Fuller (2000, 11–27) do seem to provide the means for revision, correction, and challenge of every principle (and therefore every social focus of value) they "contain," including their own tenets. Article Five of the Constitution of the United States of American and Article 135 of the Constitution of the Mexican United States are both examples of the above. If this were the case, then civic republican constitutionalism would, in effect, work as viable means to subject the organization of science to a dialectical model in which all principles, including the normative dimension, could be subject to the scrutiny of scientific practice.

Or is this the case? Aren't there *any* instances of collective value orientation that might escape or resist the possibility of scrutiny and change under such a regime? Aren't there *any* metatheoretical principles that might allow certain groups or individuals to exercise unscrutinized power under the auspices of legitimacy?

Consider, for example, the case of the determination of the franchise; this is, those allowed to have political import (i.e., those allowed to vote) in the system established by a "civic republican" constitution. For example, contemporary social and liberal democracies limit the franchise in regards to theories of nationality and nationalism. In practice, this does translate into groups of individuals exercising unscrutinized power under the auspices of legitimacy. That is, at many times authorities and other agents who are (at least theoretically) under the formal scrutiny of the constitutional powers that represent a particular franchise (i.e., the electorate of the United States) exercise authority and power over actors that are excluded from such rights (i.e., foreign nationals); all under the auspices of legitimacy.

Who will constitute the franchise of Fuller's republic of science, then, is a key question to ascertain who might be able to escape scrutiny, and thereby lack any incentive to change or challenge aspects of an arrangement from which they benefit politically.

And what about the principles of civic republican constitutionalism? As stated above, the framework they create may be able to scrutinize them. But are they, nevertheless, situated in the same location as other political principles? In other words, what implication does the fact that discussion and deliberation are ultimately guided by these principles have on the discussion and deliberation of opposing or incommensurate political positions?

In this regard, I side with Joseph Rouse (1991), who contends that the principles that guide Fuller's project of scientific constitutionalization are not placed in the same situation as the other forms of "knowledge" that are subject to the conditions of scrutiny these principles create.

The problem I identify with Fuller's, and all previous, attempts at producing an account of normativity capable of being compatible with the metaphysical commitments of a *strict* rendering of naturalism, is the insistence of developing it through the use, and under the assumptions, of a theory, a logic, or a metaphysics of *identity*, as opposed to a theory, a logic or a metaphysics of difference or *distinction* (see Luhmann 2002). Within a "logic, or theory of identity" everything that Is, Is in virtue of a set of characteristics proper to Itself: entities and phenomena are *identified* independently. These characteristics, then, may (in "naturalist" accounts) have some sort of metaphysical, or ontological relation (i.e., causality) to other "things."

Differently, in a "logic or theory of difference," metaphysical form is the result of distinction. This is, within this mode of thought everything "Is" only in regards to that which it "Is Not." Difference is metaphysically constitutive. The "world," the "Universe," or the "Absolute" is understood as an indeterminate infinity in which form results from determination (i.e., Husserl [1913] 1982). In the words of George Spencer-Brown (1969), "A universe comes into being when a space is severed or taken apart" (1969, v). One might say that a logic of difference leads to a metaphysical dialectic or sorts.

When naturalism is attempted through a theory of identity, the characteristics of the phenomenon at hand are explained in relation to the characteristics of another phenomena that are also defined by the enumeration of determined characteristics. As one climbs the causal or explanatory ladder, one is inevitably led to a final, sovereign source of Being or (epistemic) Authority that transcends all causality or explanation. This is the limit of the "dialectical" model.

I hold, then, that a normative account developed through a strict adherence to the logic of difference consistent with a strict rendering of the dialectical model (and therefore with naturalism) can be obtained.

In this sense there are various scientific attempts at the study of life and cognition that pursue a stance continuous with G. Spencer-Brown's (1969) appeal to a "theory or logic of difference." Among these one finds Heinz Von Foester's (2003) program of "second order cybernetics," Humberto Maturana and Francisco Varela's (1980) "autopoietic biology," and the radical embodied approach to the study of cognition (i.e., Chemero 2009; Thompson 2007; Varela et al. 1991). Within all this research, cognition and all cognitive phenomena, including "knowledge" is understood as a product of the cognitive agent's experience of the difference between itself and the "environment" in the sense that both parts of the equation "specify each other."

Two basic positions (found in the annals of philosophical history) emerge from these empirical studies:

1. The principle of relativism: or the position that "reality," "Truth," epistemic "validity," phenomena, and (in this particular case) cognition, are relative to the specific location, circumstances, perspective, or situation of a coordinate system (i.e., a cognitive agent).
2. The principle of constraint: historically expressed in different ways, in this particular case it refers to the view that although cognition is relative to a coordinate system (cognitive agent), it is not arbitrary: *systems are constrained to their situation.* In regards to the value that cognitive agents award to the products of offline (or abstract) cognition (i.e., imagination, theory, and hypothesis, "representation," etc.) the principle of constraint is equivalent to the position of doxastic involuntarism, or the claim that cognitive agents lack ". . . direct voluntary control over beliefs . . . and that [they] . . . have only rather weak degree of 'long range' voluntary control over (only) some of [their] beliefs" (Alston 1988: 260; see also Mark Douglas West and Miika Vähämaa's chapters in this volume).

At least two previous epistemological positions have been attempted out of that which obtains from the above described empirically motivated account of bodily situated cognition. The first, commonly labeled neurophenomenology (i.e., Thompson 2007; Varela 1996) can be understood as a universalization of the above two principles. This position leads to a transcendental phenomenology of sorts, in which all cognitive phenomena, and therefore all intellectual positions, are relative and constrained to situation, with the *exception of the claim that all cognitive phenomena is relative and constrained to situation.* The result is an ontologically naturalist version of Husserlian ([1913] 1982) phenomenology. This position however, is hardly compatible with a strict reading of the first principle (nor with strict dialectics, nor naturalism).

The second attempt, commonly labeled radical constructivism (i.e., Von Foester 2003; Von Glasserfeld 1995), is characterized by its strict commitment to the first principle, and, through it, to self-referentiality; its defining characteristic being the inclusion of the observer within their own account. In this sense, for example, Heinz Von Foester contends if one considers that observation is relative to an observer, one must also then take into account that any "description (of the universe) implies the one who observes it" (2003, 247). Through this move, they account for their specific epistemological standpoint. This is, they claim that their positions are valid for themselves, and abstain from judging the possibility for validity of their accounts in regards to any other observer (Von Foester 2003, 247–59; Maturana 1988). All validity external to the observer is then dependent on the situation of the receiver. Also called the "principle of hermeneutics" this position contends that *it is not the speaker but rather the listener who determines the meaning of a message* (Von Foerster and Porksen 1998, 100).

Two issues arise in regards to this latterly described outcome of the two aforementioned tenets. The first is that it isn't clear how it is that radical constructivism does not fall into an instance of circular arguing (again, the validity of *x* is explained by *x*). The second is that such a position seems to lead to widespread relativism that doesn't provide much ground for a normative social epistemology or political theory: from its perspective all (honest) cognitive positions seem to be equally valid.

I argue, however, that following on some, seemingly unnoticed, consequences of ultimately spatiotemporally situated (embodied) cognition, radical constructivism's problems with circular reasoning can be adequately remedied, and, furthermore, provide a normative account that leads to a strict naturalist social epistemology. The move is rather simple: as discussed above cognitive constraintment implies that the epistemic validity awarded to the products of offline cognition is beyond the volition of the agent. If this is so, then a stronger rendering of the principle of hermeneutics can be arrived at because the relationship between certainty and doubt is involuntary, and therefore (from a first person perspective epistemic) valuation is also involuntary, external validity for all cognitive phenomena (i.e., claims to "knowledge") is contingent on a receiver, *but not to his volition, rather to his involuntary cognitive situation.*

Among other things, this allows for an escape of radical constructivism's problems with circular reasoning. That is, I can claim that all the above is valid from my perspective, on account that I am (constrained) unable to find a solution to tenet (a), while at the same time I am constrained to recognize that I'm constrained (tenet [b]). I cannot claim that this situation is universal, however, for this would be incompatible with the first tenet (a). What I can do is claim without fear of dishonesty that I am unable to hypothesize a more plausible state of affairs for those that I identify as my cognitive peers (i.e., other humans). The result is that I am then constrained to hypothesize that others like me are also in (a) + (b), including you, leaving the possibility for revision of this situation. If my hypothesis is correct, however, then you will also be as constrained as I am. I can then, only request that you acknowledge this. If you are (as constrained as I am in this regard), however, then (a) + (b) has the same normative import for you as it does for me.

Finally, if you had not considered the above before this moment, then my communication has managed to alter your cognitive situation. If it has, then you have already (involuntarily) awarded your interpretation of my communication (this text) with cognitive validity. In that case, it could be said that, in regards to your specific cognitive situation before this communication took place, its results (from your perspective) constitute knowledge.

From my perspective, I am constrained to the following hypotheses that obtain from the above:

1. "Knowledge" can be understood as *communication that has the capacity to (involuntarily) alter cognitive situatedness (or limits of certainty)*, and;
2. (Natural) cognitive authority can be understood as *the capacity awarded by the cognitive situation of a determined agent to another cognitive agent, to alter this very same cognitive situation.*

Without any situation transcendent standpoints for the determination of normativity, or of cognitive authority, all cognitive validity can be said to be contingent on the capacity of cognitive agents to be changed. Without any situation transcendent standpoints for the determination of normativity of cognitive authority, all normative prescriptions that dictate certain types of communication that may lead to change as invalid, are invalid themselves.

This, then, leads to the first positive position that might lead to normative prescription, namely, that the "ideal" organization for society based on its possibilities to produce knowledge and innovation is that of a society in which all communication that has the capacity to change limits of certainty is able to do so without interference.

The specifics on how this may be achieved are beyond the scope of the present work. However, said sources of interference can be discerned from intuition aided with the results of social studies pertaining to the production of knowledge, namely (x) spatial, material, and temporal constraints to communication and (y) social hierarchical organization capable of repressing or delaying (in virtue of x) communication that might lead determined collective cognitive changes that are found at odds with the interests of the ranking members.

The resulting normative social theory is precisely one of absolute epistemic dialectics. It, in turn, obtains from an account of individual cognitive dialectics that is able to norm without removing its own position from the possibility of scrutiny, and without falling into circular reasoning. Again, the future of social epistemology lies in furthering its commitments to naturalism.

BIBLIOGRAPHY

Alston, William P. "The Deontological Conception of Epistemic Justification." *Philosophical Perspectives* 2 (1988): 257–99.

Barnes, Barry. *Interests and the Growth of Knowledge*. London: Routledge and Kegan Paul, 1977.

Chemero, Anthony. *Radical Embodied Cognitive Science*. Cambridge, MA: MIT Press, 2009.

Foester, Heinz von. *Understanding Understanding: Essays on Cybernetics and Cognition*. New York: Springer, 2003.

Foester, Heinz von and Bernhard Pörksen. *Wahrheit ist die Erfindung eines Lügners: Gespräche für Skeptiker*. Heidelberg: Carl-Auer-Systeme, 1998.

Fuller, Steve. "Epistemology Radically Naturalized: Recovering the Normative, the Experimental and the Social." In *Cognitive Models of Science*, edited by Ronald Giere, 427–59. Minneapolis: University of Minnesota Press, 1992.

Fuller, Steve. *The Governance of Science: Ideology and the Future of the Open Society*. Buckingham: Open University Press, 2000.

Fuller, Steve. *Social Epistemology*. 2nd ed. Bloomington, IN: Indiana University Press, 2002.

Fuller, Steve. *The Knowledge Book: Key Concepts in Philosophy, Science, and Culture*. Montreal and Kingston, CA: Mcgill-Queens University Press, 2007.

Fuller, Steve and James H. Collier. *Philosophy, Rhetoric, and the End of Knowledge: A New Beginning for Science and Technology Studies*. Mahwah, NJ: Lawrence Erlbaum Associates, 2004.

Glasersfeld, Ernst. *Radical Constructivism: A Way of Knowing and Learning*. London: Falmer Press, 1995.

Husserl, Edmund. *Ideas Pertaining to a Pure Phenomenology and to a Phenomenological Philosophy: First Book*. The Hague: M. Nijhoff, (1913) 1982.

Latour, Bruno and Steve Woolgar. *Laboratory Life: The Construction of Scientific Facts*. Princeton, NJ: Princeton University Press, 1986.

Locke, John and Peter Laslett. *Two Treatises of Government*. Cambridge, England: Cambridge University Press, (1689) 1988.
Luhmann, Niklas and William Rasch. *Theories of Distinction: Redescribing the Descriptions of Modernity*. Stanford, CA: Stanford University Press, 2002.
Marx, Karl. *A Contribution to the Critique of Political Economy*. New York: International Publishers, 1970.
Maturana, Humberto R. "Reality: The search for objectivity or the quest for a compelling argument." *The Irish Journal of Psychology* 9, no. 1 (1988): 25–82.
Maturana, Humberto R. and Francisco Valera. *Autopoiesis and Cognition: The Realization of the Living*. No. 42. Springer, 1980.
Putnam, Hilary. "Why Reason Can't Be Naturalized." *Synthese* 52, no. 1 (1982): 3–23.
Quine, W. V. "Epistemology Naturalized." In *Knowledge and Inquiry: Readings in Epistemology*, edited by K. Brad Wray, 245–60. Ontario: Broadview Press, 2002.
Rouse, Joseph. "Policing Knowledge: Disembodied Policy for Embodied Knowledge." *Inquiry* 34, no. 3 (1991): 353–64.
Spencer-Brown, George. *Laws of Form*. London: Allen & Unwin, 1969.
Thompson, Evan. *Mind in Life: Biology, Phenomenology, and the Sciences of Mind*. Cambridge, MA: Belknap Press of Harvard University Press, 2007.
Varela, Francisco J. "Neurophenomenology: A Methodological Remedy for the Hard Problem." *Journal of Consciousness Studies* 3, no. 4 (1996): 330–49.
Varela, Francisco J., Evan Thompson, and Eleanor Rosch. *The Embodied Mind: Cognitive Science and Human Experience*. Cambridge, MA: MIT Press, 1991.

Part IV

ENVISIONING OUR HUMAN FUTURE

Chapter 20

Visioneering Our Future

Laura Cabrera, William Davis, and Melissa Orozco

The study of knowledge within academic philosophy has to a large extent been the domain of a cadre of practitioners that write in a very particular style for a specialized and limited audience. Unsurprisingly, such practices have not fostered mainstream consumption of epistemological theories and analyses. Social Epistemology (SE), a movement aimed to challenge the traditional conception of knowledge and provide a normative agenda for Science and Technology Studies (STS) work, has recently gained more critical attention and expanded its reach in such outlets as the Social Epistemology Review and Reply Collective (SERRC). In this chapter, we emphasize how the principles of "visioneering"—the engineering of a clear vision—can help SE's original vision come to light by moving social epistemologists beyond their traditional academic roles and have a positive and stimulating impact on society at large. Questions raised within the discussions of SE and visioneering are questions that philosophy of technology (PoT) has long engaged, particularly normative ideas regarding how the public should view technological development (e.g., determinism and constructivism). Here, we promote an "un-disciplined" PoT approach—an example of SE—that speculates on and evaluates trends, developments, and concrete cases as they occur. Like SE, "un-disciplined" PoT moves from the "ideal situations" of traditional philosophy into the messy, and often unclear, real scenarios. Both SE and un-disciplined PoT call for changes regarding the practices of philosophy in our contemporary world while maintaining a critical stance regarding sociotechnical projects.

We begin by discussing how SE, as a movement, posits a new role for philosophers in its goals and objectives. We then discuss the concept of visioneering and examine how visioneering can guide SE while at the same time SE tools can be used as part of visioneering assessment. The chapter concludes by examining the parallels and possible interactions between SE and PoT and how we may imagine this project as a kind of visioneering. In particular, we posit that collaboration between these two interdisciplinary fields, when working toward a shared vision, would help identify the theoretical and normative positions of various visions of emerging technologies and facilitate a normative stance in relation to these projects, as well as foster discussion and engagement beyond academia.

SOCIAL EPISTEMOLOGY: ITS CONTEXT AND PROGRAM

At the end of the twentieth century, the emergence of several naturalistic approaches to knowledge in philosophy compelled a reimagination of the role of epistemology. One of the reasons driving epistemologists in this new direction was—in addition to Quine's theses on psychology as substitute for philosophy (Quine 1969)—the increasing number of accounts analyzing the social contexts of scientific and technological production that spread during the 1980s.[1] The empirical cases they presented for topics like objectivity or truth, topics previously considered the exclusive domain of philosophy, generated a diverse set of responses in the then nascent naturalistic epistemologies. However, all shared a similar concern: if philosophical arguments were going to impact societies, it was time for epistemologists to do more than just create models of individual knowers and their environments.

Reacting to these naturalistic approaches, Steve Fuller's vision of SE reformulated the problem of knowledge in completely new terms: not by identifying acts of knowing, but rather in terms of access.[2] This reformulation, as will be discussed later, suggested defining the role of epistemologist as policy makers: *bureaucrats* prepared to prevent the concentration of resources by some groups of the scientific community and to foster the distribution of knowledge through society. Fuller's program unfolds, accordingly, as a metatheory for current perspectives on science that still ignore the historical depth essential to treat knowledge as something that "is in the world of which it is about" (Fuller 1992). But SE, with its broad interdisciplinary horizon, also presents itself as an empirical research program for the study of science, and as a knowledge policy free of the institutional mechanisms that, at present, determine and limit the analytical scope of philosophical and sociological reflections.

These features of SE are obviously interrelated. However, in recent years, Fuller has focused more on promoting his program as a metatheory than on strategies for the normative study of science as a way to encourage a knowledge policy. Here, Fuller struggles not only in setting his account of the problem of knowledge, but also in prioritizing the role of epistemologists as policy makers in this twenty-first century. This asymmetrical development of SE seems due to the varied interpretations on the naturalization of knowledge.[3] In order to better grasp SE's purposes and integrate them in a more parallel development, we must revisit its first principles.

The question of what makes SE different from other approaches concerned with the production of knowledge in contemporary societies has been addressed in several works.[4] SE's first striking feature is its categorical resignation from the definition of knowledge of traditional epistemology and the tasks it supposes. The general reformulation of knowledge as a principle of social organization, and the consequential analysis that Fuller proposes, rejects the strict boundaries between philosophy of science and the sociology of knowledge. Thus, during the emergence of SE in the late 1980s, one of its main aims was to challenge these sharply defined boundaries that made the differences between fields appear irresoluble.

Initially, one must understand that the answer to the underlying problem of knowledge production is simply how to disseminate knowledge among society, and scientific knowledge specially requires bringing this forth. Therefore, for Fuller, the need for SE depended on the marriage between philosophy of science and sociology of knowledge (Fuller [1988] 2002). Fuller's SE thus emphasized that the question of knowledge was how to organize scientific production as a continuous growing mechanism, not as a business in which one determines the conditions of knowers to hold beliefs.

However, after more than twenty-five years of discussion in SE, this mission appears to be lost given the implicit agreements that have led philosophers and sociologists to acknowledge the need of addressing the normative dimensions of science from very similar standpoints.[5] This apparent conciliation, however, does not really focus on the problem of knowledge in the terms Fuller stated in the 1980s. Without a more definite idea of the conditions that would allow philosophers to become policy makers, SE is likely to fall prey to the same inertia it criticizes in other fields.

In this regard, Fuller only provides a sketch of the role of philosophers as policy makers. He imagines a future graduate program where epistemologists could acquire managerial and philosophical skills that help them to pursue careers as bureaucrats (Fuller 2002). Similarly, Fuller's ideal policy maker is only somewhat represented in his theses on intellectuals (Fuller 2006). For Fuller, intellectuals and social epistemologists should be critical of scientific and technological projects that escape public debate and inhibit seeing other lines of research as equally valuable as those currently developed. This kind of criticism is necessary for creating and consolidating a base for knowledge that is open to accepting errors and, therefore, is equally attentive to other possibilities.

Another model corresponding to what science policy should entail runs through Fuller's *The Governance of Science* (2000). The vision Fuller presents here offers a better perspective on how social epistemologists could translate philosophical problems into pragmatic constraints, in particular considering that scientific and technological artifacts are produced within specific economic, political, and social settings.[6] Fuller insists that the exercise of free inquiry serves as the main characteristic of science. This point suggests that anyone, regardless of one's circumstances, could be allowed to make significant contributions to the production of knowledge despite one's agreements/disagreements with the canons of the disciplines. More precisely, science cannot be taken as a collective enterprise if those who are willing to dedicate their lives to it depend invariably on the approval and whims of elites (Fuller 2000).

These general models undoubtedly offer grounds for criticism and improvements. However, in spite of Fuller's somewhat vague outline regarding the role for policy makers, he clearly opposes the propagation of epistemologists' curricula ensnared in armchair reflection and its antagonistic shift as mere spectators of scientific, lab-based activities. Yet, in the complex dynamic of the study of science and technology before and after the emergence of SE, philosophers have been hesitant to direct their attention to science and technology policy. Philosophers of science in particular have confined the limits of their practices to the discussion of problems without arguing how their solutions would fit in the current state of affairs. Thus, they rarely cross the line that could lead to policy advice. The causes that deviated philosophy of science's comprehensive vision from cultural and social matters have a complicated historical background that helps explain their isolation in the realm of science and technology policy.[7] For SE to take this mission back, it must analyze why the transition to science policy from philosophy still lags behind on important issues.

The connection between philosophy of science and science policy became a concern shortly before the emergence of SE (Campbell 1984). Donald Campbell and Patrick Suppes, as philosophers and administrators for institutions like the National Science Foundation, reflected on how philosophy of science could become a sort of new critical foundation to advise science policy. They offer accounts of why philosophy of science should inform policy makers, the weight public policy issues should have in the discipline, and the kinds of philosophical models that would serve best to analyze them. In their accounts, they stress the benefits of using the naturalistic approach to intervene in some scientific areas along with the need for a major theoretical development that eventually would cover other policy issues.

These points illustrate the potential benefits that philosophy could bring for science policy but from a perspective in which the problem of knowledge does not require a full reorganization of the structure of the disciplines. Consequently, the results do not question the logic that informs science policy. This does not mean, however, that the normative recommendations hypothesized by philosophers lack their own forms of control. As any other administrative tool, these recommendations can include their own forms of evaluation to confirm the congruence of the philosophical analysis. Yet, as we have suggested, SE encourages one to identify the historical threads that permeate current motivations around the search for knowledge. So, aside from an influence in Analytical Social Epistemology, which has followed a similar path but devising its own approach, the introduction to science policy of these endeavors just portrays the possibilities of cooperating with policy advisors in an exploratory way.

Engaging fully in policy making supposes the recognition—by institutions and individuals—of science and technology as social activities. Likewise, the recognition of epistemologists as experts in managing the production of knowledge cannot just be asserted. If we consider Fuller's program of SE in its three different—but integrative—directions: as a metatheory of science studies; as an empirical research program for the study of knowledge production; and as knowledge policy (Fuller 1992), then SE's main goals have yet to be reached. Social epistemologists, overall, have not entered into the field of policy making. This basic remark has inspired some of Fuller's followers to truly put his vision into practice.[8] In the publish-or-perish environments that academics and researchers at universities currently face, applied forms of SE represent a clear move away from the common practices of traditional epistemology to a piecemeal intervention in knowledge production.

Reenvisioning the job of epistemologists in contemporary times has acquired more relevance within SE and beyond, yet, as the scenario of contemporary research changes, taking on a more complex dynamic, many challenges are likely to arise. Wittkower et al. (2014) address part of this problem when they analyze the purposes of PoT in relation to the incentives philosophers get for extending their reflections to the public. One of the many lessons in their analysis for our discussion of the future roles perspective opened by SE, is the firm conviction that philosophers can effectively address several issues of public interest. This closer interaction with the public would necessitate a reexamination of the credentials of philosophers (beyond writing peer-reviewed articles for academic journals) and the eventual significance of this interaction can help to connect deeply with research interests and theoretical motivations. Reenvisioning the job of epistemologists is, therefore, not simply a question of "updating" current practices in the sense of just adding new information to what we supposedly already know. Putting forth an effective vision of SE that can be translated into a practice requires a discussion of the progress of SE in academia in terms that account for the perils of detaching its exercise from the public. This is not just a trivial issue, since creating a solid bridge between academic programs and their "demand" in society is not regarded here as a mere business venture, but as a way to revolutionize the basis of contemporary society. In short, the issue of how exactly those inspired by SE could defend their affiliation to science and technology policy opens up questions regarding the way philosophy and sociology could inform and intervene beyond their traditional academic domains.

Remedios (2003), Valero (2006), and Collin (2013) have already taken steps to address these concerns. Besides revisiting some of the debates Fuller has initiated, they have also confirmed, with different emphases in their analyses, the ambitiousness of Fuller's normative program. They point out that SE needs more than theoretical development in order to create firmer grounds for its practice. Practitioners also must decide how to make SE's approach more compatible with neighboring disciplines and the public in general. Each review has offered remarks that call for a more modest approach than what Fuller has outlined. Remedios, Valero, and Collin agree that even if a political legitimation of knowledge is feasible, the success of SE in democratizing science cannot be reduced to some sort of menial check and balance exercise since an effective knowledge policy should investigate the social function of science. Fuller's SE needs a more global appraisal that enables relevance to other fields given that the systematization in philosophy still fails to accomplish its goals. In that regard, aside from the answers that are still pending in the philosophical domain, the task of identifying and critiquing the social forces and influences responsible for knowledge production that SE has advanced is an interconnected topic for shaping shared visions of the future. SE not only has theoretical significance, inasmuch as it defends a central role in the knowledge-forming process; it can also develop an as-yet unrealized practical importance.

VISIONEERING AND ITS RELATION TO SOCIAL EPISTEMOLOGY

The neologism "visioneering" encompasses two main concepts: that of the "visionary" on the one hand, and that of the "engineer" on the other. When these two concepts are combined, they embody the hybrid nature of visioneers, that is, individuals or groups of individuals that actively engineer a clear vision they have about the future (Kim and Oki 2011; McCray 2012). This active engineering involves the skillful direction and creative application of scientific and technological principles to the development of novel processes, structures, and equipment, as well as social institutions, movements, and frameworks.

Visioneers go beyond having visions about the future. They participate in a diverse set of activities to shape the particular worldview they envision.

Patrick McCray has suggested that visioneering entails "developing a broad and comprehensive vision for how the future might be radically changed by technology, doing research to advance this vision, and promoting one's ideas to the public and policy makers in the hopes of generating attention and perhaps even realization" (2012, 13). A common example portrays the technology enthusiast as undertaking a varied course of future-directed activities based on his or her technological visions about the future. Raymond Kurzweil and Eric Drexler both fit this common idea of the visioneer. Kurzweil, an inventor, futurist, author, and the current director of engineering at Google, not only holds radical and positive technological visions about the future as portrayed in his books and talks[9]—the visionary aspect that is an essential part of the visioneering motivation—but he also actively engages in shaping the future he envisions—the engineering aspect. As an example of the latter, he cofounded in 2008 Singularity University, a new brand of university whose mission is "to educate, inspire and empower leaders to apply exponential technologies to address humanity's grand challenges."[10] Singularity University not only locates the visions of their cofounders at the forefront of the curricula, but it also serves as a site for fostering active engineering through exponential technologies to positively shape the future (Cabrera 2014). Eric Drexler, an MIT-trained engineer, advances a specific view about nanotechnology, namely one with a focus on molecular engineering (Drexler 1986). His vision includes self-replicating nanoscale machines capable of assembling materials and devices as needed, or that could replace molecules and cells in the human body. Both Kurzweil and Drexler envision the technologies and applications they promote as paths to transforming individuals and, thus, society. They promote radical uses of emergent technologies that aim to enable humans to master their bodies and environments in unprecedented ways, ultimately overcoming current biological limits.

The type of changes promoted by visioneers require such profound transformations of the status quo that not having the right social, political, and economic environments can prevent a visioneer's plans from flourishing. Thus, having a vision about a changed society (whether by radical technological change or fundamental changes in our social institutions) is not enough; it is also often required that visioneers actively build communities and networks that enable them to connect their ideas to other members of society, in particular people that could help them create the right environment to fulfill their vision. This approach is implicit in policy making and can be regarded as a soft version of social engineering.

While the main focus of visioneering so far has been on radical technological futures, visioneering is not limited to those types of visions. Moreover, as Alfred Nordmann claims, "visioneering can also be viewed as a widespread collective activity" (2013, 89; cf. Cabrera 2014). For instance, individuals can exploit their capacity to organize discourses and disseminate them via powerful symbolic technologies, a particular position that academics via their universities and languages could definitely do more to promote. Visioneering in this sense can also generate new forms of institutions or social manifestations. Cabrera (2014) has taken this idea a step forward by arguing that visioneering need not be something left to technology enthusiasts (in the sense of engineering physical artifacts)[11] but rather is something academics and the general public can and should become involved with more directly. Thus, a visioneer need not be an engineer in the sense of Drexler or Kurzweil. He or she must, however, have a clear normative vision regarding how developing technologies can significantly change humanity, as well as the drive to pursue it passionately toward its completion. In this sense, people who passionately and courageously pursue their vision can also be considered visioneers.[12]

People can engage in visioneering in other ways. For instance, they can assess current visioneering projects, or challenge and scrutinize current visions of the future. Decisions about the type of future we want to live in should not be left to elites or experts in one field or another (Cabrera 2014). Cabrera has emphasized the need to promote more engagement, from other fields of academia to the general public, to determine what different groups, cultures, and societies wish for the future and how these diverse goals might best be realized. In this regard, visioneering, either as actively engineering the future or as part of its assessment, allows us to defend the democratic ideal (e.g., by creating spaces to participate and collaborate in visioneering projects) that allows the growth and diversification of today's technological, political, social, and cultural ecosystems.

Visioneering assessment, as has been understood within STS, deals with the peculiar qualities of visions and their scientific, as well as social, political, and philosophical, plausibility (Grunwald 2007; Ferrari et al. 2012). It also concerns itself with implicit conceptions of the good life, with desirability and acceptability, and even with broader perspectives concerning our common future and social development (Grunwald 2007; Ferrari et al. 2012). STS scholarship, however, often lacks the normative aims that we as social epistemologists count as fundamental to our projects. SE moves beyond descriptive accounts and toward more explicit engagement with such philosophical questions as what technologies, institutions, structures, and mind-sets will lead to "improved" lives. SE can contribute substantially to this kind of visioneering, in particular by advocating critical attention to the social, philosophical, and normative dimensions underlying the types of visions (whether technological or not) proposed by visioneers. Thus, SE serves as way to engage in forward-looking projects. At the same time, it provides a normative empirical framework for assessing the knowledge, values, assumptions, and desires driving these visions.

There are other ways in which visioneering and SE interact, and that is that visioneering can be seen as a framework to move from the visions that gave rise to SE, namely the recovery of the normative in the organization of inquiry (Fuller 2014), to visioneering them. Visions of the future of SE have spread from reflections on the field's twenty-fifth anniversary. These reflections include the acknowledgment that room still exists for recovering the initial normative framework of SE. Fuller himself recognizes that the normative program he established has made very few advances in the last twenty-five years (Fuller 2014). The problems and tensions that appear when one tries to define the job of social epistemologists in terms of the visions underlying the field led us to consider the aspects of visioneering that SE could borrow to give practical significance to its visions.

Visioneering, we argue, serves as an example of the type of SE—the organization, management, and development of knowledge—advocated by Fuller (2002). Understood this way, social epistemologists like Fuller who directly engage with the shaping of future institutions, research agendas, epistemic communities, and public policies can be regarded as visioneers. Fuller's vision of SE has specific goals about what SE should be and how it should be carried out, and he has actively worked to develop the frameworks and institutions for such a project. One example of this has been his key role, along with founder Jim Collier, in setting up the SERRC.[13] However, even though Fuller has actively built communities and networks that enable him to connect his ideas to other members of the academic world, his visions lack the same widespread impact that the techno-focused visions of Kurzweil or Drexler have garnered. In this context, the remaining questions entail how to ensure that other forms of visioneering flourish in ways similar to those of Kurzweil and Drexler. More importantly, can we ensure that other forms of visioneering do not end up committing to the technologically deterministic and often teleological views that seem to be part of the visioneering practices carried out by technological enthusiasts (McCray 2012)?[14]

It may be important to clarify that even if visioneering is largely a new concept whose boundaries are still being defined—inasmuch as it encourages academics and the public to generate and collectivize their visions—it aligns with SE's interests in democratizing scientific and technical knowledge. Next, a common concern that runs in both are the historical bases that are used to shape the visions of the future. Is history a resource from which we could pick up lessons? If so, how? But, what is less evident here is the way the engineering aspect of visioneering could be developed as a "nontechnological" response. Furthermore, there are important questions regarding how we could counterbalance techno-futuristic visions that certainly imply unnecessary risks and a biased perspective whose realization implies deep changes in society, with other type of visions.

VISIONEERING SOCIAL EPISTEMOLOGY: USING KEY ASPECTS OF PHILOSOPHY OF TECHNOLOGY AND ANTICIPATORY GOVERNANCE

Techno-visioneers, such as Kurzweil or Drexler, often present their views as being beyond human control: technologies will change people, environments, and societies, so we should simply develop the right technologies. They propose a form of "hands-off" view that allows "technologies to run their course"—as if technologies have a teleology and will of their own, ignoring the social, political, and economic factors that affect technological development and integration. For instance, Patrick McCray's (2012) work describing visioneers as "enthusiasts" is an adequate description, but it is also insufficient. By choosing the label "enthusiast," rather than "determinist," McCray utilizes a rhetorical strategy of concealing the normative agendas of the visioneers he describes.

These accounts encourage a lapse into a malaise where humans think they can do little to affect states of affairs, and they emphasize the alienation found in technological determinist texts (Heidegger 1979; Ellul 1962; Marcuse 1991). In this regard, un-disciplined philosophers of technology emphasize the ethical, social, and political co-constructions of technologies in order to invite more public participation in such projects, challenging deterministic views of technological development.[15] Like SE, un-disciplined PoT assesses trends and developments that result from the adoption of particular technologies and makes normative claims regarding how future development should occur. Taking this into consideration, the specific form of visioneering we advocate below incorporates an "un-disciplined" PoT perspective (Davis 2013) that engages lay publics and is critical of deterministic views.

Particular challenges for visioneering involve speculating about and governing the future effects of emerging technologies. Technologies appear easier to govern in the early phases of development and implementation. However, the social, economic, and political impacts of such technologies resist early identification and tend to be realized only after technologies have permeated societies and cultures.[16] Unfortunately, attempting to control and regulate technologies after they enter the marketplace becomes so laborious and economically arduous that efforts to do so strain resources (Johnston 1984). Speculative ethics and anticipatory governance offer a tentative solution: explore and evaluate, in advance, the values, politics, and social goals that societies want their technologies to promote.

Recent accounts of anticipatory governance (Fuller 2009, 2010; Johnson 2011; Guston 2008, 2014; Sarewitz 2011) provide valuable alternatives and spaces for "un-disciplined" philosophers of technology to enter and help shape visioneering debates. David Guston, for instance, defines anticipatory governance as "a broad-based capacity extended through society that can act on a variety of inputs to manage emerging knowledge-based technologies while such management is still possible" (2008, vi). In a similar vein, Fuller argues that anticipatory governance amounts to a "strategy to facilitate the acceptance of new technosciences by inviting people to voice their hopes and concerns in focus groups, science cafes, and computer-based interactive spaces before the innovations are actually implemented" (Fuller 2009, 209).

Approaches to governance of scientific and technological developments emphasize nonlinear and co-constructive flexibility (Ozdemir et al. 2009), avoid top-down policy pronouncements (Groves et al. 2010), encourage public participation (Quay 2010), and promote responsible innovation (Owen et al. 2012). However, important questions about organizing and implementing anticipatory governance remain. By leveraging users, lay publics, and experts (academic or otherwise), an un-disciplined PoT offers a step toward resolving these issues.

Philosophical inquiry into "the good life" in this century demands that the sciences and technologies purportedly improving our lives and our world match the values our societies wish to promote and explain the changes that should be made (Michelfelder 2011). Such demands, however, require scholars to eschew "end of the day assessment"[17] and instead offer speculative judgments about the kinds of human-technology relationships that will permit democratic and participatory intervention once emerging technologies leave design spaces and enter into the complex social, economic, and political frameworks of our daily lives.

Speculative ethics and anticipatory governance projects—exercises in visioneering—should not ignore deterministic views, but these should be tempered with social constructivist accounts that illustrate how, over time, human interpretations of function and utility serve to stabilize and close debate (Pinch and Bijker 1984). Constructivists examine a multitude of factors (social, political, economic, environmental, etc.) that contribute to how competing technological designs eventually stabilize as well as extrapolate the means by which to exert greater control over technological design and implementation in the future, revealing that technologies do not develop autonomously. Constructivism provides more than an alternative to deterministic views of technology. It also points the way for general publics to become involved in shaping the design and reception of technologies, promoting a kind of activism that empowers those in the relevant social groups to advocate for change—a project SE endorses. People and institutions play important roles in determining which technologies we develop and why, and leveraging these roles enables various actors to participate in anticipatory governance of emerging technologies.

As SE emphasizes, visioneering projects should include more public input at, and even prior to, the design phases of emerging technologies. It is here where visioneering can use key aspects of un-disciplined philosophers of technology and anticipatory governance, in assessing the visions behind emerging technologies, including incorporating values into the technological designs themselves. Free from the "ideal situations" and strictures of academic philosophical writing, un-disciplined PoT is attuned to the social, ethical, and political dimensions of technologies. Such commentary is necessary to reinvigorate public discussions of technologies' impact on the human condition and promote active engagement in anticipatory governance of visioneering projects.

If visioneering aims to be more than an academic, scientific, and engineering affair, we must reimagine the roles and interactions of lay publics with developers and policy creators. These interactions could involve presenting descriptions and normative implications of emerging technologies to lay publics, or the creation of truly interactive arenas where publics and experts can exchange ideas and make recommendations for future developments. Here we would like to highlight two current venues in which these interactions could develop, one within academia and the other without. The first consist of university classrooms, and centers for research, like Arizona State University's Center for Nanotechnology in Society (CNS). An informed lay public must be able to comprehend the ethical, legal, and social implications/applications of technologies. University classrooms and research centers, as well as informative opinion pieces like those from think tanks,[18] serve as ideal incubators for such education, but they do not reach enough audiences. The second arena consists of outlets like focus groups and computer-based interactive spaces—for instance, online forums that not only present emerging technological developments but also invite feedback from readers. The *MIT Technology Review* and *Future Tense* take steps in this direction as they offer commentary on developing technologies and offer spaces for lay publics to respond. Simply offering a comments section, however, is insufficient for actually making spaces for lay publics to interact with developers of emerging technologies and scholars working in anticipatory governance.

VISIONEERING AS ANTICIPATORY GOVERNANCE

Promoters and critics of visioneering projects, must present technological development as co-constructed—context-dependent and coproduced (dependent on both the technologies and societies that create and implement them)—for lay publics to take seriously their roles in anticipatory governance. Technological determinism presents a picture of a world controlled by technology, not by individuals and societies, and that image leads to complacency and inaction. The vision of SE focusing on more public engagement implies that such activity can impact those technically oriented and elite actors who develop technologies and write law and policy. A lack of public engagement expands the gap between makers and users, enabling a form of social determinism where elite actors control the laws and policies of emerging technologies.

Peter-Paul Verbeek's (2005) call for philosophers of technology to work with designers of technologies and technological systems marks a step toward infusing philosophical insight into the design process. Exploring the ethical dimensions of the proposed technologies, however, does not fall only to philosophers, scientists, and/or engineers. Designers of technologies cannot accurately predict the uses of those same technologies (Feenberg 1995), thus, more philosophical input in the design phases of technologies has the potential to provide more accuracy. Anticipatory governance, as a key aspect of visioneering projects, should provoke publics to consider how these technological visions may alter our world and ourselves.

In order to better understand the implications of such changes and goals, the general public should have a grasp of the normative foundations and goals of those attempting to "visioneer" future societies. Historical perspectives are crucial to such understanding, like that offered by McCray (2012), but equally important are analyses and critiques of these ventures. Human-technology relations must be explained and understood as vital to emerging technology projects in order for lay publics to see their roles in co-constituting technological change. A way beyond the deterministic and teleological current coursing through the works of many visioneers involves public debate and discussion of the ethical, economic, environmental, political, and social theories found at the base of emerging technology projects because these schemes will, as the constructivists remind us, impact multiple societies and cultures, laws, and policies.

To encourage more public participation in visioneering as anticipatory governance, lay publics need explications of the potential positive and negative effects of such ventures. Those who publish outside of expensive (and largely unread) academic journals and presses, such as un-disciplined philosophers of technology, reach broader audiences than their academic counterparts. This wider reach permits them to, potentially, influence greater numbers of lay publics, a potential that many philosophers of technology covet (Wittkower et al. 2014). Much could be learned from their rhetorical maneuvering—the outlets they publish their work in, their style of writing, how they engage various audiences, and their ability, particularly in the case of Morozov (2013), to influence perspectives on cutting-edge developments. An un-disciplined PoT, then, is not meant only for philosophers, traditional, or otherwise, and that means *how* philosophers of technology communicate and *whom* they engage takes on ever-greater importance. These aspects of un-disciplined philosophers of technology to engage audiences outside of academia and motivate them to participate in deliberations regarding the course of future developments can be used in visioneering the future of SE.

Visioneering could certainly benefit from different actors espousing a variety of normative positions regarding technological developments in order to promote a more ecological deliberation over the kinds of futures they wish to inhabit. These normative agendas should be fully elucidated in a manner that is comprehensible and clear to more than just academic audiences in order to avoid top-down styles of governance (Groves et al. 2010). Building on the work of Diane Michelfelder (2011) we argue that involving different publics in assessing the normative claims and implications of visions can allow for important values to emerge that might otherwise go unheeded, open up avenues that might otherwise go unnoticed for reframing and recasting issues, as well as permit questions to be raised that might otherwise go unvoiced (Michelfelder 2011, 58).

Visioneering assessment, similar to speculative ethics, provides lay publics with an opportunity to evaluate the potential impacts such technologies will—and, importantly, already *do*—have on values like privacy and autonomy and the laws governing them.[19] As Rebecca Roache (2008) observes, "Attending to possible future scenarios is an important part of addressing current issues" (320). When considering our future, we must also take stock of our present. Speculative and anticipatory thought regarding emerging technologies would not simply serve as a plan for evaluating specific future technological developments. It would also permit engagement with broader ethical questions regarding how humans and our environments currently exist and the types of human-technology engagement we wish to cultivate in the future. Attention to these values now may enable us to better plan for future contingencies that arise from adopting emerging technologies.

A tension clearly still exists: multiple perspectives may not prove any more adept at predicting future outcomes than those of the elite actors. However, by focusing on the core values that our current and future technologies exhibit permits reflection on, for instance, the type of surveillance we do and do not want and the ways in which our technologies promote and inhibit such surveillance (Wittes and Chong 2014).

Like Michelfelder (2011), feminist ethicists and epistemologists (Jaggar 1991) invite us to pay attention to those groups that have not been permitted to voice their opinions and situations, and lay publics regarding emerging technologies constitute one type of underrepresented group that demand to be heard if we truly wish for the technologies, laws, and policies that we visioneer to be representative of the people they serve.

CONCLUSION

SE is a boundary-breaking program that has contributed to understanding the naturalization of knowledge by emphasizing the consequences of our conceptions of knowledge in the real world, as well as in the always-problematic stratification of the disciplines. The challenge of imagining the future of SE, or, for that matter, explaining the roles of social epistemologists in that future, remains an ongoing task. Nevertheless, we recommend that social epistemologists, as they explore the ethical, legal, and social implications of different visions of the future, take more active visioneering roles by engaging in the accomplishment of their own visions. In doing so, they can use approaches from un-discipline PoT and anticipatory governance. Visioneers must give precise accounts regarding how their visions contribute in meaningful ways to societies and how their visions will transform societies, economics, environments, and values. Unlike techno-visioneers, whose visions are permeated with technologically deterministic views, and who generally have positive expectations on technology bringing better futures, progress, and development in their respective fields, visioneering within SE takes an antideterministic view showing how humans and technologies co-construct ideas of progress and the futures these projects will entail.

In order to gain legitimacy and a place within academia and outside of it, SE must leverage the interdisciplinary backgrounds of its practitioners to provide critical commentary about science, technology, and society. Spanning three countries, multiple cultures, and two languages, the authors of this chapter have, by way of regular, distance correspondence, and discussions, attempted to bring our various perspectives and training to bear on the specific topic of visioneering. We have each approached this topic with particular normative inclinations developed from our experiences as graduate students and early professionals in disparate fields of study, yet our goal has been to craft a plausible (and in our view desirable) path for the practice of SE in visioneering that transcends our unique viewpoints. If visioneering is to be an interdisciplinary project, then we offer this chapter as evidence of the potentials such collaborative work entails.

NOTES

1. Main works that represent this tradition are and became the foundations for a larger bibliography: Bloor (1991), Latour and Woolgar (1995), and Lynch (1985). A compilation where these main positions and other works are contrasted is offered in Pickering (1992).

2. We will return to the problem of access at the end of the chapter when examining visioneering as a form of anticipatory governance of emerging technologies.

3. Remedios (2003) offers an account on how Fuller's argument challenges other approaches to naturalism.

4. Remedios (2003), Valero (2006), and Collin (2013) have addressed this question.

5. For a discussion on this, see Rouse (2003) and Remedios (2003).

6. We will return to this point in later sections while describing social constructivist views of technological development.

7. Reich (2005) has made a suggestive number of clarifications regarding the works of the Vienna Circle as a movement, whose interest went beyond the project of unity of science. His account on how diverse concerns disappeared as positivism spread in the United States, excluding political and cultural issues from the methods and theories of philosophy of science, is evocative when we consider the acceptance of Fuller's SE in sociological and philosophical studies of science.

8. For instance, Thomas Basbøll has explored applied forms of SE, turning the practice of social epistemology into a writing consultant service (Basbøll 2012).

9. Within his authored books reflecting his visions about the future, it is worth mentioning *The Age of Spiritual Machines* (1999) and *The Singularity is Near* (2005).

10. Singularity University (SU) is not a traditional university where students matriculate and graduate; rather, the model is one in which for a period of two months a graduate program is run in which students attend talks by appropriate experts about different converging technologies, do site visits to core places in the region of Silicon Valley, and engage in a project that aims to bring together their learning outcomes with visions of how to positively change the world. singularityu.org/.

11. We understand definitions of technology to include social institutions, methods, and institutionalized habits of thought (cf. Jasanoff 2003).

12. Not all visions are shared, giving rise to outcomes that are disapproved by society. In this regard the term visioneer is morally neutral, as it does not imply that visioneers' acts are moral per se.

13. Steve Fuller's (1988, 2002) *Social Epistemology* not only provides critiques of current knowledge systems and philosophies, but also sketches curricula and agendas for the practice of SE.

14. Steve Fuller's own work, most recently in *Preparing for Life in Humanity 2.0* and *The Proactionary Imperative: A Foundation for Transhumanism* (coauthored with Veronika Lipinksa), also display both technologically deterministic and teleological tendencies. On his view, scientific and technological advances drive our societies, and we should accept this notion and work to implement it in the best manners possible (Fuller 2013; Fuller and Lipinska 2014).

15. Other main themes in PoT over the last century and a half include technological utopianism, dystopian futures, and a mix of wary enthusiasm and criticism.

16. The Collingridge (1980, 11) dilemma explicates such difficulties.

17. We are, in part, reacting to the place and role of philosophy as described by Hegel:

> Philosophy, as the thought of the world, does not appear until reality has completed its formative process, and made itself ready. History thus corroborates the teaching of the conception that only in the maturity of reality does the ideal

appear as counterpart to the real, apprehends the real world in its substance, and shapes it into an intellectual kingdom. . . . The owl of Minerva, takes its flight only when the shades of night are gathering.

G. W. F. Hegel (1991), *Elements of the Philosophy of Right*, Preface.

18. Cf. Benjamin Wittes and Jane Chong's (2014), "Our Cyborg Future: Law and Policy Implications."

19. For example, Wittes and Chong (2014) note how many of our current technologies, like Google Glass, make us agents of and subject to surveillance (18). Even a Web browser, by recording our searches, page views, and clicks, make our private thoughts potentially public and searchable by third-party applications.

BIBLIOGRAPHY

Basbøll, Thomas. "The Supplementary Cleck: Social Epistemology as a Vocation." *Social Epistemology* 26, nos 3–4 (2012): 435–51.

Brey, Philip. "Philosophy of Technology After the Empirical Turn." *Techne* 14, no. 1 (2010): 36–48.

Brey, Philip. "Anticipatory Ethics for Emerging Technologies." *NanoEthics* 6, no. 1 (2012): 1–13.

Cabrera, Laura Y. "Visioneering and the Active Role of Engagement and Assessment." *NanoEthics* 8, no. 2 (2014): 201–206.

Campbell, Donald. "Science Policy from a Naturalistic Sociological Epistemology." In *PSA: Proceedings of the Biennial Meeting of the Philosophy of Science Association*, 14–29. Chicago: University of Chicago Press, 1984.

Coeckelbergh, Mark. "Hacking Feenberg." *Symploke* 20, nos 1–2 (2012): 327–30.

Coenen, Christopher and Elena Simakova. "STS Policy Interactions, Technology Assessment and the Governance of Technovisionary Sciences." *Science, Technology & Innovation Studies* 9, no. 2 (2013): 3–20.

Collin, Finn. "Two Kinds of Social Epistemology." *Social Epistemology Review and Reply Collective* 2, no. 8 (2013): 79–104.

Collingridge, David. *The Social Control of Technology*. London: Frances Pinter, 1980.

Davis, William. "Philosophy of Technology 'Un-Disciplined.'" *Social Epistemology Review and Reply Collective* 2, no. 12 (2013): 12–13.

Drexler, Eric. *Engines of Creation*. New York: Anchor Books, 1986.

Ellul, Jacques. "The Technological Order." *Technology and Culture* 4, no. 2 (1962): 394–421.

Ellul, Jacques. *The Technological Society*. Translated by John Wilkinson. New York: Vintage Books, 1964.

Feenberg, Andrew. *Alternative Modernity*. Berkeley: University of California Press, 1995.

Feenberg, Andrew. *Transforming Technology*. Oxford: Oxford University Press, 2002.

Ferrari, Arianna, Christopher C. Coenen, and Armin A. Grunwald. "Visions and Ethics in Current Discourse on Human Enhancement." *NanoEthics* 6, no. 3 (2012): 215–29.

Fuller, Steve. "Epistemology Radically Naturalized: Recovering the Normative, the Experimental, and the Social." In *Cognitive Models of Science: Minnesota Studies in the Philosophy of Science*, vol. 15, edited by Ron Giere, 427–59. Minneapolis: University of Minnesota Press, 1992.

Fuller, Steve. *The Governance of Science*. Buckingham and Philadelphia: Open University Press, 2000.

Fuller, Steve. "¿Se Han Extraviado Los Estudios de la Ciencia en la Trama Kuhniana? Sobre el Regreso de los Paradigmas a los Movimientos." In *Desafíos y Tensiones Actuales en Ciencia, Tecnología y Sociedad*, edited by Cerezo Ibarra, 71–98. España: Biblioteca Nueva, 2001.

Fuller, Steve. *Social Epistemology*. 2nd ed. Bloomington, IN: University of Indiana Press, 2002.

Fuller, Steve. *The Intellectual*. Cambridge: Icon Books, 2005.

Fuller, Steve. "Review of the Handbook of Science and Technology Studies." *Isis* 100, no. 1 (2009): 207–209.

Fuller Steve. "The New Behemoth (Review of Acting in an Uncertain World)." *Contemporary Sociology* 39, no. 5 (2010): 533–36.

Fuller, Steve. *Preparing for Life in Humanity 2.0*. London: Palgrave MacMillan, 2013.

Fuller, Steve. "Social Epistemology: The Future of an Unfulfilled Promise." *Social Epistemology Review and Reply Collective* 3, no. 7 (2014): 29–37.

Fuller, Steve and Veronika Lipinska. *The Proactionary Imperative*. London: Palgrave MacMillan, 2014.

Groves, Christopher Robert, Robert Lee, Lorraine Frater and Gavin David J. Harper. "CSR in the UK Nanotechnology Industry: Attitudes and Prospects." Working Paper Series 57, 1 July. Cardiff: Centre for Business Relationships, Accountability, Sustainability & Society, 2010.

Grunwald, Armin. "Converging Technologies: Visions, Increased Contingencies of the Conditio Humana, and Search for Orientation." *Futures* 39, no. 4 (2007): 380–92.

Guston, David H. "Preface." In *The Yearbook of Nanotechnology in Society: Presenting Futures*, vol. 1, edited by Erik Fisher, Cynthia Selin, and Jameson Wetmore, v–viii. New York: Springer, 2008.

Guston, David H. "Understanding 'Anticipatory Governance.'" *Social Studies of Science* 44, no. 2 (2014): 218–42.

Heidegger, Martin. *The Question Concerning Technology and Other Essays*. Edited by William Lovitt. New York: Garland Publishing, Inc, 1979.

Jaggar, Alison M. "Feminist Ethics: An Introduction." *Forum: A Women's Studies Periodical* 16, no. 1 (1991).

Jasanoff, Sheila. "Technologies of Humility: Citizen Participation in Governing Science." *Minerva* 41, no. 3 (2003): 223–44.

Johnson, Deborah G. "Software Agents, Anticipatory Ethics, and Accountability." In *The Growing Gap Between Emerging Technologies and Legal-Ethical Oversight*, edited by Gary E. Marchant, Braden R. Allenby, and Joseph R. Herkert, 61–76. New York: Springer Science+Business B.V., 2011.

Johnston, Ron. "Controlling Technology: An Issue for the Social Studies of Science." *Social Studies of Science* 14, no. 1 (1984): 97–113.
Kelly, Kevin. *What Technology Wants*. Viking: Penguin Group, 2010.
Kim, Joon and Taikan Oki. "Visioneering: An Essential Framework in Sustainability Science." *Sustainability Science* 6, no. 2 (2011): 247–51.
Kurzweil, Ray. *The Age of Spiritual Machines*. New York: Penguin Books, 1999.
Kurzweil, Ray. *The Singularity is Near*. New York: Penguin Books, 2005.
Lanier, Jaron. *You Are Not a Gadget*. Vintage: Random House, Inc, 2011.
Latour, Bruno and Steve Woolgar. *La Vida en el Laboratorio*. Madrid: Alianza, 1995.
Lynch, Michael. *Art and Artifact in Laboratory Science*. London: Routledge and Kegan Paul, 1985.
Marcuse, Herbert. *One-Dimensional Man*. Boston, MA: Beacon Press, 1991.
McCray, W. Patrick. *The Visioneers*. Princeton, NJ: Princeton University Press, 2012.
Michelfelder, Diane P. "Dirty Hands, Speculative Minds, and Smart Machines." *Philosophy and Technology* 24, no. 1 (2011): 55–68.
Morozov, Evgeny. *The Net Delusion*. New York: Penguin Books, 2012.
Morozov, Evgeny. "The Internet Ideology: Why We are Allowed to Hate Silicon Valley." *Frankfurter Allgemeine Zeitung* (2013). www.faz.net/aktuell/feuilleton/debatten/the-internet-ideology-why-we-are-allowed-to-hate-silicon-valley-12658406.html.
Mumford, Lewis. "Authoritarian and Democratic Technics." *Technology and Culture* 5, no. 1 (1964): 1–8.
Nordmann, Alfred. "Visioneering Assessment: On the Construction of Tunnel Visions for Technovisionary Research and Policy." *Science Technology and Innovation Studies* 9, no. 2 (2013): 89–94.
Owen Richard, Phil Macnaghten, and Jack Stilgoe. "Responsible Research and Innovation: From Science in Society to Science for Society, with Society." *Science and Public Policy* 39, no. 6 (2012): 751–760.
Ozdemir Vural, Don Husereau, S. Hyland, S. Samper, and Mohd Zuki Salleh. "Personalized Medicine Beyond Genomics: New Technologies, Global Health Diplomacy and Anticipatory Governance." *Current Pharmacogenomics and Personalized Medicine* 7, no. 4 (2009): 225–30.
Pacotti, Sheldon. "We Are the Builders of Tech Revolutions. Why Are They Still a Surprise?" *Paul Writer* (2014). paulwriter.com/builders-tech-revolutions-still-surprise/.
Pickering, Andrew, ed. *Science as Practice and Culture*. Chicago and London: University of Chicago Press, 1992.
Pinch, Trevor and Wiebe Bijker. "The Social Construction of Facts and Artifacts: Or How the Sociology of Science and the Sociology of Technology Might Benefit Each Other." *Social Studies of Science* 14, no. 3 (1984): 399–441.
Quay Ray. "Anticipatory Governance: A Tool for Climate Change Adaptation." *Journal of the American Planning Association* 76, no. 4 (2010): 496–511.
Quine, W. V. O. *Ontological Relativity and Other Essays*. New York: Columbia University Press, 1969.

Reich, George. *How the Cold War Transformed Philosophy of Science: To the Icy Slopes of Logic*. Cambridge, UK: Cambridge University Press, 2005.

Remedios, Francis. *Legitimizing Scientific Knowledge: An Introduction to Steve Fuller's Social Epistemology*. Lanham, MD: Lexington Books, 2003.

Roache, Rebecca. "Ethics, Speculation and Values." *NanoEthics* 2, no. 3 (2008): 317–27.

Rouse, Joseph. "Remedios and Fuller on Normativity and Science." *Philosophy of the Social Sciences* 33, no. 4 (2003): 464–71.

Sarewitz, Daniel. "Does Climate Change Knowledge Really Matter?" *Wiley Interdisciplinary Reviews: Climate Change* 2, no. 4 (2011): 475–81.

Suppes, Patrick. "Philosophy of Science and Public Policy." In *Proceedings of the Biennial Meeting of the Philosophy of Science Association*, vol. 2, edited by P. D. Asquith and Philip Kitcher, 3–13. Chicago: University of Chicago Press, 1986.

Valero Lumbreras, Angel. *Epistemología Social y Política del Conocimiento: Un Análisis del Programa de Investigación de Steve Fuller*. Tesis de Doctorado. Madrid: Universidad Complutense, 2006.

Verbeek, Peter Paul. *What Things Do*. Translated by Robert P. Crease. University Park, PA: Pennsylvania State University Press, 2005.

Verbeek, Peter Paul. *Moralizing Technology*. Chicago: The University of Chicago Press, 2011.

Verbeek, Peter Paul. "Expanding Mediation Theory." *Foundations of Science* 17, no. 4 (2012): 391–95.

Wittes, Benjamin and Jane Chong. *Our Cyborg Future: Law and Policy Implications*. The Brookings Institution (2014). www.brookings.edu/research/reports2/2014/09/cyborg-future-law-policy-implications.

Wittkower, D. E., Evan Selinger, and Lucinda Rush. "Public Philosophy of Technology." *Techné* 17, no. 2 (2014): 179–200.

Chapter 21

Dreaming the Future

What It Means to be Human

Emma Craddock

Imagination is the starting point.[1] It can act as a springboard for knowledge and the seed for political transformation. Indeed, as Solnit contends, "Politics arises out of the spread of ideas and the shaping of imaginations" (2005, 35). The boundless nature of imagination provides us with the freedom to explore potential futures and endows us with a sense of possibility; there is not only one possible future, but many. I intend to explore some of these possibilities here, considering the role of imagination and its relationship with politics. In particular, I will be drawing on ideas concerning dreams, stories, hope, and the future, situating a political project of imagination within the overarching theme of what it means to be human.

Imagination is not just about what *will* be known but, importantly, what *could* be known. This sense of possibility provides the context within which we can dream and wonder about the future, and perhaps most significantly, hope for a better future. However, given that such imagining points us in the direction of the unknown it also encourages uncertainty, a state which is unsettling for many. For we not only have the power to dream about a better future but also to imagine nightmares of a worse one. As Virginia Woolf stated six months into World War I, "The future is dark." And yet, she follows this by claiming that it is "the best thing the future can be, I think." As Solnit affirms, Woolf does not mean "terrible" when she uses the word dark here but merely inscrutable and uncertain. Such uncertainty is not necessarily negative; it can "become the grounds for hope" (2005, 29) and hope is central to imagining a better future.

In fact, Braidotti contends that "hope is a way of dreaming up possible futures: an anticipatory virtue that permeates our lives and activates them" (2013, 192). Here Braidotti draws our attention to the process of dreaming about the future. The dream stretches the realms of possibility; it plays with the boundaries of what is real and epitomizes fantasy. However, while dreams have the potential, and the appeal, of being limitless and fantastical, they nevertheless contain the seeds of tangible realities. As Subcomandante Marcos of the Zapatista Army states, "In our dreams we have seen another world. . . . And this new, true world was not a dream from the past; it was not something that came from our ancestors. It came to us from the future; it was the next step that we had to take" (*Zapatistas!* 1994).

However, hope is not prognostication, it is not about predicting future outcomes but is "an orientation of the spirit, an orientation of the heart; it transcends the world that is immediately experienced and is anchored somewhere beyond its horizons" (Solnit 2005, 13–14). Clearly then such an emotion and state is tied closely to imagination and the possibilities of dreams. It is also about being able to imagine alternatives that may not currently seem viable and believing in the potential for these to be realized at some point, perhaps in the more distant future. However, while hope deals with the wide landscape of possible futures, we should not make the mistake of believing that it is a vague, idealistic notion without grounding or tangibility. In fact, Braidotti's comment draws attention to the *active* process of hoping; she describes hope as something which "activates" our lives. Likewise, Solnit contends that "hope calls for action; action is impossible without hope" (2005, 14). Here hope is the motivation for political action, it "should shove you out the door" to protest for transformation (5).

Moreover, while hope may be the driving force behind collective action for change, like dreams it begins in the imagination of an individual. Indeed, a now famous call for change begins from the starting point of "*I* have a dream" (Martin Luther King). However, it is not only the actions and speeches of charismatic leaders that we should pay attention to but also the "countless small actions of unknown people" (Chomsky 2012, 23). For radical change and ideas happen and begin at the margins, not the center. Yet, too often our eyes are trained on the "main stage" of politics when we should be looking, as Solnit (2005) phrases it, "in the dark," away from the glaring lights shining on the center stage.

In this volume (and see 2013), Victoria Peake writes about her visualization of the trans/posthuman that took center stage of an art exhibit, encouraging the audience to imagine the possible futures of "the human." This idea of enhancing and perfecting "the human" is a project of development; it assumes that history is a linear and forward-marching process that builds upon that which has come before. Furthermore, it is not only a forward march but an upward one too, aiming to improve and progress humanity. Yet, if we shift our gaze to the sidelines we will see the presence of another, radically different narrative that breaks with such a linear structure and embraces the messiness and complexities that this brings. As Subcomandante Marco's statement hints, such a narrative entangles the past, present, and future in the same way that dreams seldom follow a straightforward route but instead turn the world upside down. Rather than following a linear trajectory, it is slowly being recognized and proclaimed by individuals and movements at the margins that "history is not an army. It is a crab scuttling sideways, a drip of soft water wearing away stone, an earthquake breaking centuries of tension" (Solnit 2005, 4). To visualize history as an army, marching forward and upward, invokes ideas of a vanguard who leads such a task, with the rest following in the ranks. But people are breaking ranks.

The Zapatista Army of National Liberation is a key example of this new imaginative politics. They emerged in Chiapas in the early 1990s to defend their land rights against encroaching capitalist forces. But in their wake they inspired an alternative, creative form of political resistance that has resonated around the world. As the *Notes from Nowhere* collective exclaims, the Zapatistas "understood the power of subjectivity, spoke the language of dreams, not just economies" (2005, 23). Furthermore, they "translated the struggle into a language that the world can feel, and invited us all to read ourselves into the story, not as supporters, but as participants" (2005, 23–24). There are several key points to make here; firstly, emphasis is placed on a common *feeling* that is shared across cultures and space drawing our attention to the centrality and force of human emotions. Secondly, the movement encourages participation rather than mere observation; individuals are enticed to step out from the shadows at the sidelines and take part rather than passively watch as the drama unfolds on the main stage. Finally, the movement sparks creativity and highlights the ways in which we, as humans, make sense of our experiences and the world around us through narrative. Indeed, the story is a powerful tool that draws on the imagination, as Polletta argues, "People *do* things with stories. They entertain and persuade, build social bonds and break them, make sense of their worlds and, in the process, create those worlds" (2006, 14). Creativity is central to interpreting, constructing, and retelling our experiences of reality. In fact, Duncombe remarks: "It is not that reality doesn't exist—it is more that by itself it doesn't really matter. Reality is always refracted through the imagination, and it is through our imagination that we live our lives" (2007, 18).

The traditional narrative of human progress toward perfection has been the dominant story throughout history, echoed in religions from thousands of years ago and modern analyses of the future technological advances and potentials of the (post) human. But such a story trains the eyes on the endpoint, it encourages us to look forward (and upward) toward the end goal of human perfection, viewing history as a progressive, linear march toward this apex. In contrast, movements like the Zapatistas and the activists that they inspire are challenging this traditional focus on the ends and instead concentrating on the means. In fact, a dangerous risk of focusing on the distant future is that we develop an escapist attitude that neglects the present and severs roots with reality. Instead, by focusing on the *process*, such movements ensure that they act in accordance with their values while forging spaces in the present where the future is imagined. Prefigurative politics demonstrate that another world is possible, bringing these future possibilities into the present in a particular place for a short period of time.

In fact, inspired by the Zapatistas, the "autonomous geographies" project signals one example of how alternative futures can be imagined and enacted in the present. This represents spaces where people desire to build a "non-capitalist, egalitarian and solidaristic" mode of political, social, and economic organization and engagement (Pickerill and Chatterton 2006, 730). Such projects "interrupt space and time . . . to open up perspectives on what might be" (Pinder 2002, 229). Significantly, "autonomous geographies" is a politics of hope focused on imagining future possibilities in the present everyday lives of individuals and, by doing so, make such possibilities appear more probable. Indeed, *Notes from Everywhere* refers to such places as "spaces of hope" (2003, 29). Moreover, Chatterton (2006) admits that such a politics of imagining capitalist alternatives is a messy, experimental, and complex process but embraces this dynamism. In a similar manner, the "chaos" of carnival has been invoked in protests to embody this spirit of rebellion, festival, and fun. Yet, on a deeper level, it has historically reflected the inversion of social order, a day where the world is turned upside down (Bakhtin 1984). It is when the dreamworld and reality meet, the possibilities are endless and the imagination is free to roam. As the Situationist slogan proclaims, "Be realistic: demand the impossible." Indeed, the Situationists celebrated creativity, creating "situations" that momentarily shattered the status quo and importantly, encouraged individuals to open their imaginations and think of the possibilities.

Furthermore, there is a break from singular grand narratives that dominate the landscape and an emergence of a scattering of stories, a plurality of voices at the margins that are dreaming and speaking of an alternative way of doing politics. In fact, *Notes from Everywhere* contends:

> People were sick of sacrificing themselves for the sake of gigantic game plans which didn't account for their individual needs, their humanity, their culture, their creativity. They were unwilling to be soldiers or martyrs in movements whose big-picture and top-down solutions were to be imposed on the "masses," which too often existed only in the imaginations of vanguard revolutionaries. (2005, 23)

Here we are introduced to the idea of the people's imagination, as individuals and as an alternative collective. Importantly, the people do not equate to "the masses," these are not faceless numbers in a crowd but individuals with ideas, dreams, and imaginations that come together as a collective to negotiate their visions for the future. I am reminded here of Hardt and Negri's concept of "the multitude," a global network of individuals where differences can be expressed openly and freely while still achieving common ground. The task at hand, then, is to collectively produce (rather than discover) "*the common*" that allows individuals to act together without negating their differences (Hardt and Negri 2005, xiv). Indeed, Hardt and Negri contend that "democracy is a matter not only of formal structures and relations but also of social contents, how we relate to each other and how we produce together" (2005, 94). Here, emphasis is placed on human relationships and interaction along with the need for collective creation, again highlighting the centrality of the creative imagination. Furthermore (and perhaps most significantly here), the multitude is a project of *possibility*. At the same time, such possibilities are grounded in the everyday lives and experiences of individuals (as explored by the autonomous geographies project and evident in local social and community centers throughout the United Kingdom).

It is not accidental that "creativity" is listed along with "humanity" in the quotation above; indeed, such creativity is a distinctly human characteristic. However, what do we mean when we appeal for a politics that accounts for humanity? At first glance, such a statement makes little sense—surely politics is by its very nature human? Yet, if we think back to the origins of the Zapatista struggle—the need to fight to retain rights to their land against the attack of capitalist forces, we edge closer to the crux of the matter and the kernel of this alternatively imagined politics. Center-stage politics has forgotten humanity. In the contemporary context of advanced capitalism and neoliberal values the focus has shifted away from the human and further onto profit. It is an often heard gripe nowadays that politics is run by big business; the puppet masters behind the scenes (increasingly becoming less hidden) are the corporations who fund and manoeuver the political parties on stage. Certainly, such politics does not forget the human entirely; problematically and worryingly it privileges a certain kind of individual who has the power and resources to make money and accumulate more and more in an endless process. However, it is particularly good at forgetting the human casualties that occur during this process, those who are ignored or stepped upon, pushed to the sidelines, into the dark where it is hoped no one will see them and that they will be forgotten. What the Zapatistas and the movements that followed did is raise their voices and demand to be listened to. They actively stepped out from the darkness and proclaimed that there is an alternative, and the first step toward it is imagining it. In the process they bring our attention back to the central question of what it means to be human.

So far I have discussed the qualities of creativity, imagination, and hope, arguing that such qualities are key threads woven into the rich tapestry of humanity. Indeed, the very question "what does it mean to be human?" demonstrates the prominence of meaning to humanity. All of these qualities tie into emotion—we imagine a better future that satisfies our wants and desires, to hope for such is a deep-felt human emotion, and we express both our imaginings and our hope through creativity. Indeed, at the heart of the human condition is emotion. As Eve Ensler (2011) defiantly exclaims, "I am an emotional creature." To have compassion is to be "humane," alternatively we describe those who lack empathy and emotion as "monsters"; we strip them of their human status. And yet, as Jasper and Polletta muse, "Somehow, academic observers have managed to ignore the swirl of passions all around them" (Goodwin 2001, 1). The problem is that the emotional has traditionally been depicted as the antithesis of the rational. And it is reason which "takes center stage. It is reason, after all, that distinguishes us as human" (Duncombe 2007, 11). That which speaks to the emotions (or the heart and body, which have typically symbolized the emotional as opposed to the mind) is dangerous. Like imagination, emotion is uncontainable. It is messy and threatens to overspill into areas of life such as the political which (according to Enlightenment ideals) should be a matter of the discerning mind and concerned with reason.

But relying on reason alone and forgetting emotions is equally (if not more) dangerous. It results in men in business suits (produced by the slave labor of children thousands of miles away), sitting in a boardroom calculating algorithms to produce the most profit possible while failing to take into account the *human* cost of their decisions and actions. I have mentioned that to be compassionate is to be humane, and it is this principal emotion that I wish to focus on here. For such alternative movements as the Zapatistas and those influenced by them reveal a key motivation at the core of such imaginative political possibilities, namely empathy. In fact, while their struggles are mainly rooted in local concerns, such groups form part of a wider global network that links up individual struggles and connects one to another through a sense of shared struggle and a common humanity. Like Hardt and Negri's multitude, despite and in fact *because* of their differences, individuals can seek to find a common ground rooted in this shared humanity and empathy for other human beings around the world. It is about caring, for others who you may never have met and never will meet, and also for the consequences of your actions. Significantly, this empathy has grown to include nonhuman animals and nature within its scope. In contrast, the traditional story of human progression toward perfection places the human at the center, as the star of the story, an approach that reflects the dominant tale of consumer capitalism. In their imaginings of an alternative future, movements at the margins reject this capitalist storyline in favor of a more holistic and sustainable one which not only takes humanity into account but also considers its impacts on the environment and grounds this in a politics of empathy. In this alternative story we should be considering and *considerate* of the present and future impact of our actions and choices on the environment, on other human beings, and on future generations, starting with today.

Such an approach may seem optimistic, but isn't that the beauty of imagination? Having the ability to dream for an alternative, better, future and visualizing its realization. Furthermore, as I have pointed out throughout, such a dream is not floating around aimlessly in the air; it is grounded in practical, present demonstrations of an imagined alternative politics and reflects the voices of many. For a long time we have been told (and have retold) the same old story not only of human progress, but also of humanity. We have repeatedly heard that human beings are inherently selfish, unruly, or even evil. Therefore, humans must be constrained (whether by a social contract, authority, or other means) in order to encourage harmony between individuals and to suppress this "natural" urge to dominate and/or do harm. But alongside this dominant narrative is a lesser heard alternative one that proclaims the opposite, that human beings inherently possess the qualities for mutual aid and voluntary cooperation (see, e.g., Kropotkin 2009). Maybe it's time to change the narrative? Or, instead, perhaps we should accept that there is a plurality of stories that exist, many that are deliberately pushed to the margins, and we should turn our ears to listen to more than just the dominant one.

Finally, while I have focused in this chapter on events that some would (arguably) say belong to a past struggle from which it is time to move on, I contend that the waves made by the Zapatistas are still being felt around the world. In fact, at the time of writing, we are experiencing austerity measures imposed by the UK government in response to the global financial crisis and the damaging consequences these measures have on individuals in our society. If you shift your eyes to the margins you will see movements of individuals rising up against this, inspired by the creativity and imagination of these older movements and motivated by empathy. And if history is water wearing away a stone (Solnit 2005), then perhaps this is another droplet. Indeed, it is easy to ignore such actions that occur away from the main stage but, like Solnit (2005), I contend that we need to learn to look in the shadows and who knows, maybe one day we will be able to see in the dark.

NOTE

1. "The power of imagination is infinite."—John Muir.

BIBLIOGRAPHY

Bakhtin, Mikhail. *Rabelais and His World.* Indiana: Indiana University Press, 1984.

Braidotti, Rosi. *The Posthuman.* Cambridge: Polity Press, 2013.

Chatterton, Paul. "So What Does it Mean to be Anti-capitalist? Conversations with Activists from Urban Social Centres." *Urban Studies* 47, no. 6 (2006): 1205–24.

Chomsky, Noam. *Occupy.* Brooklyn: Occupied Media Pamphlet Series | Zuccotti Park Press, 2012.

Duncombe, Stephen. *Dream: Re-imagining Progressive Politics in an Age of Fantasy.* New York: The New Press, 2007.

Ensler, Eve. *I Am An Emotional Creature: The Secret Life of Girls Around the World.* New York: Random House, 2011.

Goodwin, Jeff, James M. Jasper, and Francesca Polletta, eds. *Passionate Politics: Emotions and Social Movements.* Chicago: The University of Chicago Press, 2001.

Hardt, Michael and Antonio Negri. *Multitude.* London: Penguin Books, 2005.

Kropotkin, Peter. *Mutual Aid: A Factor of Evolution.* New York: Cosmo Pub, 2009.

Notes from Nowhere, ed. *We Are Everywhere: The Irresistible Rise of Global Anticapitalism.* London: Verso, 2005.

Peake, Victoria. "The Value of Imagining the Trans/Posthuman." *Social Epistemology Review and Reply Collective* 2, no. 11 (2013): 13–15.

Pickerill, Jenny and Paul Chatterton. "Notes Towards Autonomous Geographies: Creation, Resistance and Self-Management as Survival Tactics." *Progress in Human Geography* 30, no. 6 (2006): 730–46.

Pinder, David. "In Defense of Utopian Urbanism: Imagining Cities after the 'End of Utopia.'" *Goegrafiska Annaler Series B: Human Geography* 84, nos 3–4 (2002): 229–41.

Polletta, Francesca. *It Was Like a Fever: Storytelling in Protest and Politics.* Chicago: University of Chicago Press, 2006.

Solnit, Rebecca. *Hope in the Dark: The Untold History of People Power.* Edinburgh: Cannongate Books, 2005.

Zapatistas! Documents of the New Mexican Revolution. Brooklyn, New York: Autonomedia 1994. lanic.utexas.edu/project/Zapatistas/.

Chapter 22

Human Enhancement

Visual Representation and the Production of Knowledge

Victoria Peake

This chapter explores the debate surrounding human enhancement technologies and the changing notion of what it means to be human. It argues that imagination is vital to encourage and guide progress and innovation, and furthermore that it is important to visualize and create visual representations from such imaginations. In this way it becomes possible to orient attitudes toward such a vision, and share discursive ideas to create a shared vision of the future. Furthermore, the chapter explores the extent to which visual representation can produce knowledge by using an example from my own research, in which I have painted and exhibited a depiction of a cyborg, a human-robot hybrid, in order to tap into the discourse of human enhancement technologies, demonstrate the value of visual representation in reaching peoples' preconceptions about such possibilities, and the extent to which the visual alone has the power to produce knowledge. In order to do this viewer responses have been recorded through voice recordings and questionnaires.

HUMAN ENHANCEMENT: A NEW HUMANITY?

With scientific and technological intervention it is becoming increasingly possible to enhance our minds and bodies beyond what is considered natural (Meulen 2009). We are in a new period of human technology, including biotechnology, neuroscience, computing, pharmacology, and nanotechnology, which have the potential to improve human performance beyond what is required for good health (Miller and Wilsdon 2006a). Technologies may help us to "think better, feel happier," "improve physical skills," or "extend lifespan" (Meulen 2009, 38). As such, science and technology are enabling us to escape the limits of the human condition, thus the notion of what it means to be human is changing.

It can be argued that humanity is undergoing a radical transformation toward a post-humanity. Garreau argues that transhumanism, the term commonly applied to the cyborg strand of posthumanism is "the enhancement of human intellectual, physical and emotional capabilities, the elimination of disease and unnecessary suffering and the dramatic extension of lifespan" (Garreau 2006, 231). Furthermore, it is the "belief in the engineered evolution of 'post-humans,'" humans whose capabilities are no longer "unambiguously human by our current standards" (Garreau 2006, 232). Thus, transhumans are those that are in the process of becoming post-human (Wolfe 2010).

There are many advocates of human enhancement such as Nick Bostrum and Aubrey de Grey, while others such as Francis Fukuyama (2009) argue that transhumanism is "the world's most dangerous idea." Aubrey de Grey argues that aging is much like a disease in that it kills people despite us having the scientific and technological capabilities to slow it down or prevent it (Aubrey de Grey cited in Miller and Wilsdon 2006b). Some argue that there is nothing immoral about wanting to improve oneself since this is human nature and the nature of progression (Caplan 2006). As Ruud ter Meulen points out, "Most emerging biotechnologies are being developed in order to cure people of severe diseases," but many of them also go beyond therapy and function for the enhancement of human capacities (Meulen 2009, 38).

This, somewhat blurred distinction, between therapy and enhancement is one of the key debates within transhumanism. Some argue that it is moral and acceptable to utilize science and technology in order to restore good health, but not to be better than healthy. However, this is problematic since the distinction between therapy and enhancement is not always easy to see. For example, Oscar Pistorius, the South African double below-the-knee amputee who competed in the London 2012 Olympics, was originally refused permission to compete by the world governing body for track and field due to the "clear competitive advantage" that his prosthetic racing legs gave him (Casert 2008 and International Amateur Athletic Federation 2008; Wolbring 2008, 117). After this claim was rejected, Pistorius competed against biological-leg runners on the basis that it would maintain a level playing field (Wolbring 2008). Some argue that human enhancement technologies would lead to the treatment of the nonenhanced as "limited, defective and in need of constant improvement" and that "the rat race for ever-increasing abilities will continue with no end in sight" (Wolbring 2008, 121).

However, it is inevitable that science and technology will continue to increase possibilities and that society will progress at an exponential rate. Thus, institutions such as the Singularity University founded by Kurzweil and Diamandi in 2008 help to apply these exponential technologies in order to address humanity's challenges and positively impact the future (Cabrera 2013). It is important to accept that change is inevitable and that we must utilize and manage it in the best way possible. Such changes and ideas start with imagination and visualization.

THE VALUE OF IMAGINING: VISUALIZING THE TRANS/POSTHUMAN

During such a period of transformation and social change, it is reasonable to argue that imagination plays a key role is shaping the future. As Craddock points out in citing Duncombe (2007, 18), "Reality is always refracted through imagination and it is through imagination that we live our lives." It can be argued that by imagining what a posthuman might look like, we can orient our attitudes toward such a possibility and we can better access the moral, ethical, and political issues surrounding human enhancement technologies. Therefore, if we visually represent such imaginations it becomes a shared vision, one which has the power to evoke emotive reactions. As Harper (2002) points out, "Images evoke deeper elements of human consciousness" and a different kind of information from the written word (13). Furthermore, "Artworks can evoke a reflexive, cognitive response from viewing publics, in which powerful emotion and recognition can combine to extraordinary effect" (Eyerman and Ring 1998, 283). Therefore, it is incredibly valuable to not only imagine but also create visual ideas that can be shared in order to shape the future they imagine.

There have been many examples of art that have incorporated the transhumanist ideology, particularly bioart, which pushes the limits of what is considered biologically possible. As Miah argues, "Bioartists have begun to infiltrate scientific laboratories in the name of new expression and new knowledge," exploring "new media through which to create art that will change our way of seeing the world" (Miah 2011). In this way bioartists "conduct sociologies of the future" through adopting "technologies that are at the margins of human experience," including stem cells, cosmetic surgery, and biotechnology, emphasizing "collaboration and shared vision" (Miah 2011). Bioart, then, blurs the distinction between art and life (Mitchell 2010) shifting possibilities from the imagination into reality. Such artists have not only imagined but also actively engaged "in a diverse set of activities to shape the future they envision," such that bioartists are also "visioneer's" (Cabrera 2013, 1).

Furthermore, "In a post-human, digital age where we can easily imagine ourselves beyond the physical confines of our own mind, body and world, art can offer the tools we need to think differently about ourselves" (Stubbs and

Sillars 2008, 16). Through imagination and visual representation it becomes possible to "visioneer" our future and to "nudge society toward expansive scenarios of the technological future they imagine" (McCray 2012, 152). It is important to point out, as Cabrera argues, that "visioneering should not be an activity solely for technological optimistic entrepreneur's, but also an activity in which academics from the humanities and social sciences are also engaged" in order to avoid a future with a narrow set of visions (Cabrera 2013, 3).

THE EXHIBITION

Emphasizing the importance of an interdisciplinary approach to visioneering, as a sociologist I have conducted my own research using visual representation, exploring the value of the method. A painting was created of a woman as a human-robot hybrid, and was exhibited in public spaces in order to explore the extent to which visual representation does evoke emotive responses, whether it proves a useful tool for reaching opinions on the subject, and the extent to which visual representation can produce knowledge. Viewer responses were recorded through voice recordings of initial reactions to the piece, and through questionnaires.

The viewing sample consisted of a combination of academics, students, and staff, and nonacademics, as the exhibition took place in the atrium of a students' union, and in the sociology department common room. The study in the atrium took place on an open day at the university, such that the sample was varied, including many parents and potential students, undergraduate and postgraduate students all from a variety of subjects, and various staff from the university. This meant that the viewers had different degrees of knowledge about human enhancement, many of which stated that they had no prior knowledge of the subject.

It is important to stress that the painting was exhibited with no title or explanation, such that the viewers were not told what the painting was depicting, or of any artist's intentions or opinions. In this way, it was possible to explore the extent to which the painting required an explanation to make sense, and to isolate the knowledge obtained from the visual alone. Responses were collected through twenty-six questionnaires and voice recordings. Questions were asked regarding viewer interpretations of what the painting depicts and how the painting makes them feel, and after initial responses were recorded, viewers were asked how they felt about the idea of enhancement. The questions were intended to capture the understanding and knowledge of human enhancement prior to the viewing of the painting, and subsequent viewer understanding and feelings toward the piece.

Figure 22.1 Woman as a Human-Robot Hybrid

The first point to be made is that viewers noted the usefulness of visually presenting difficult ideas such as human enhancement. One respondent said "It is useful to convey difficult concepts to the general public as it actively makes people think," while another said, "Art can help generate discussions and bring issues into popular culture." This emphasizes the importance of having a shared vision and not limiting possible futures by restricting debates on the subject to solely academia and particular academic disciplines. By sharing ideas from the imagination through visual depictions, ideas and discussion can permeate. As Stubbs and Sillars argue, artwork has the "power to encapsulate some of the most complex debates of our times" and many artists aim "to carve out freer spaces for debate and discourse, where difficult conversations are allowed and indeed encouraged" (Stubbs and Sillars 2008, 15–16).

EMOTIVE RESPONSES

Respondents seemed to have strong reactions to the painting regarding how it made them feel, suggesting that art does indeed "evoke a reflexive, cognitive response from viewing publics, in which powerful emotion and recognition can combine to extraordinary effect" (Eyerman and Ring 1998, 283). Some negative responses included words such as "unnecessary," "dangerous," "problematic," "scary," or "pointless," while those who reacted positively used words such as "emancipatory" and "exciting" and indicated curiosity and interest about the idea. Many respondents were deeply engaged with the painting; one respondent, when asked how it made them feel, answered, "a bit naked, frankly, but in a somewhat detached way." Another said, "I am really touched by this picture; it looks like she is talking to me." Furthermore, another respondent, when asked what he thought the painting depicted, said that he could "feel it more than [he] could say it."

It seems reasonable then to argue that visual representation is a powerful tool for reaching opinions and feelings toward an idea. As Chaplin (1994) argues in reference to Becker (1981), the "aesthetic quality of an image can give an analysis immediacy, vividness, poignancy . . . in a way that can be ideologically hard-hitting" (277). In this case, the visual was effective in inducing engaged emotive responses, which fuelled greater discussion and viewers were able to quickly orient their attitudes from such emotions, thus enhancing the capacity of the researcher to capture people's preconceptions about this future of uncertainty.

PRODUCTION OF KNOWLEDGE

Furthermore, there were many viewers who had no prior knowledge of transhumanism or human enhancement technologies other than examples from popular culture such as Oscar Pistorius. However, despite this, with no explanation from the researcher, viewers mentioned some of the main arguments from academia written on this subject, such as the distinction between therapy and enhancement and the fear of disparity in society should such enhancements become readily available.

Many viewers felt positive toward enhancement for repair or for health restoration while taking the opinion that any enhancement beyond this was immoral or unnecessary. For example, one respondent said, "If it is to overcome disease, then I'm all for it, but I don't approve of science for the sake of science." Another said, "To save life it's good but just to enhance for enhancement's sake is pointless." This distinction between therapy and enhancement, however, is becoming increasingly blurred. The pursuits of technology for enhancement and for health maintenance are not so dissimilar. Medical intervention aspires to improve; many examples of such interventions are difficult to categorize. For example, cosmetic surgery is a vast field in which medicine is used for both repair and enhancement, from skin grafts to face lifts, breast implants to liposuction. Furthermore, medical science has addressed health care to aim for prevention before cure, such as screening and treating patients before they are diagnosed with an illness, increasing their health span and potentially their lifespan. Therefore, the distinction between therapy and enhancement, despite forming arguments against human enhancement technologies, is becoming increasingly blurred. Having said that, viewers noted this distinction, a debate raised in much of the literature on transhumanism, with no prompting from the researcher and no prior knowledge of the subject. This could suggest that the visual has the power to produce knowledge without an explanation.

In the majority of cases where this ethical distinction between therapy and enhancement were mentioned, the reason for taking a negative view was based on worries about disparity in society, unequal competition, and disadvantage. One respondent said, "It makes me feel uneasy because I feel it can only ultimately disadvantage the majority by making 'normal' become 'inadequate.'" Another respondent said,

> I think we should focus on the fact that the overwhelming majority of humans live in the kind of conditions that rich, white westerners would consider abhorrent for their own pets. If science and technology were interested in making better humans it should start by addressing this . . . and not worry about making life even more comfortable for the already super-privileged.

These responses reflect fear that enhancing oneself beyond ones biological capabilities would cause disadvantage through uneven distribution of opportunity and access to such possibilities.

SO IMAGINE, VISUALIZE, SHARE

In a changing world in which science and technology are creating new possibilities, imagination is key to creating and shaping innovation. To reiterate, "Reality is always refracted through imagination and it is through imagination that we live our lives" (Duncombe 2007, 18). Moreover, it is essential that we go beyond imagination to visualize and create representations of new ideas and possibilities. In this way, visions of the future can be shared in a discursive space in which debate is allowed and encouraged. As Cabrera argues, in order to "visioneer" and shape future societies to "radically transform the human condition," we need to advance the vision and "promote those ideas to the public and policy makers" (Cabrera 2013, 1). This becomes more possible when we widen the arena in which these visions are discussed, "where diverse interests and values meet" (Cabrera 2013, 1). Furthermore, emphasizing the usefulness of the visual method, it can be argued from the research that visual representation is a powerful tool for producing knowledge and gaining emotive responses to such ideas. Thus, "Artworks can evoke a reflexive, cognitive response from viewing publics, in which powerful emotion and recognition can combine to extraordinary effect" (Eyerman and Ring 1998, 283). Therefore, I argue that society *needs* to imagine future possibilities, visualize what such possibilities could look like, including the ways in which we can "visioneer ourselves" (Cabrera 2013, 2), and share these ideas in order to guide society through this rapid transformation and social change.

BIBLIOGRAPHY

Becker, Howard, ed. *Exploring Society Photographically.* Block Gallery: Northwestern University, 1981.

Cabrera, Laura. "Visioneering and Our Common Future." *Social Epistemology Review and Reply Collective* 2, no. 10 (2013): 1–3.

Casert, Raf. "IAAF Rules Amputee Oscar Pistorius Ineligible for Olympics." *New York Times.* Monday, January 14, 2008. http://www.nytimes.com/2008/01/14/sports/14iht-athletics14.9191533.html.

Caplan, Arthur. "Is it Wrong to Try to Improve Human Nature?" In *Better Humans? The Politics of Human Enhancement and Life Extension*, edited by Paul Miller and James Wilsdon, 31–39. London: Demos Collection 21, 2006.

Chaplin, Elizabeth. *Sociology and Visual Representation.* London: Routledge, 1994.

Craddock, Emma. "Dreaming the Future: What it Means to be Human." *Social Epistemology Review and Reply Collective* 3, no. 3 (2014): 52–54.

Duncombe, Stephen. *Dream: Re-imagining Progressive Politics in an Age of Fantasy.* New York: The New Press, 2007.

Eyerman, Ron and Magnus Ring. "Towards a New Sociology of Art Worlds: Bringing Meaning Back In." *Acta Sociologica* 41, no. 3 (1998): 277–83.

Fukuyama, Francis. "Transhumanism." *Foreign Policy* (2009). foreignpolicy.com/2009/10/23/transhumanism/.

Garreau, Joel. *Radical Evolution.* New York: Broadway Books, 2006.

Harper, Douglas. "Talking about Pictures: A Case for Photo Elicitation." *Visual Studies* 17, no. 1 (2002): 13–26.

McCray, Patrick. *The Visioneers: How a Group of Elite Scientists Pursued Space Colonies, Nanotechnologies, and a Limitless Future.* Princeton: NJ: Princeton University Press, 2012.

Meulen, Ruud Ter. "Will Human Enhancement Make Us Better? Ethical Reflections on the Enhancement of Human Capacities by Means of Biomedical Technologies." In *Human Futures: Art in an Age of Uncertainty*, edited by Andy Miah, 38–47. Liverpool: Liverpool University Press, 2009.

Miah, Andy. "Biotechnology, Bioethics and Bioart." *Future: Content, It's Nice That.* London, 2011. www.andymiah.net/2011/06/04/biotechnology-bioethics-bioart/.

Miller, Paul and James Wilsdon. "Stronger, Longer, Smarter, Faster." In *Better Humans? The Politics of Human Enhancement and Life Extension*, edited by Paul Miller and James Wilsdon, 13–27. London: Demos Collection 21, 2006a.

Miller, Paul and James Wilsdon. "The Man Who Wants to Live Forever." In *Better Humans? The Politics of Human Enhancement and Life Extension*, edited by Paul Miller and James Wilsdon, 52–58. London: Demos Collection 21, 2006b.

Mitchell, Robert E. *Bioart and the Vitality of Media.* Seattle: University of Washington Press, 2010.

Stubbs, M. and L. Sillars. "Life in the 21st Century—Practice Based Research and the Application of Art." In *Human Futures: Art in an Age of Uncertainty*, edited by A. Miah, 11–4 of Introductions. Liverpool: Liverpool University Press, 2008.

Wolbring, Gregor. "One World, One Olympics: Governing Human Ability, Ableism and Disablism in an Era of Bodily Enhancements." In *Human Futures: Art in an Age of Uncertainty*, edited by Andy Miah, 115–25. Liverpool: Liverpool University Press, 2008.

Wolfe, Cary. *What is Post-Humanism?* Minneapolis: University of Minnesota Press, 2010.

Chapter 23

Is Transhumanism Gendered? The Road from Haraway

Steve Fuller and Veronika Lipinska

When Donna Haraway first published the "Cyborg Manifesto" in *Socialist Review* in 1985, she thought she was extending the feminist project by using the image of the cyborg from science fiction as a model for a conscious—even sexual—being that nevertheless lacked stable ontological markers. Haraway put forward *Blade Runner*'s high-grade replicant Rachel (played by Sean Young)—deeply confused yet endearing—as the paradigm case of the "cyborg feminist" (Haraway 1991, 178). In this she had history behind her, since in its original incarnation, Alan Turing's famed test, now taken to be about telling the difference between a human and a machine, was about deciding whether the interlocutor was a man or a woman (Genova 1994). But equally for this reason, Haraway flagged, in the very paragraph of her article, that her appeal to the cyborg was "ironic." The problem is that her fans have read her manifesto in increasingly literal terms, thereby rendering her a prophet of contemporary transhumanism, an attribution that she has come to vehemently reject.

A feminist reading Haraway in 1985 would have taken her to be contesting what had passed as "radical feminism" for the previous decade or more, namely, a gender essentialism, which associated with particular ways of knowing and being with the genital arrangement of one's birth. Feminists who held such views often advocated self-segregating lesbian communities as the only way to escape male oppression (Daly 1978). And even those who did not advocate such extreme solutions nevertheless were stuck on the idea that there is a particular "women's way of knowing" that derives from, say, the experience of motherhood (Ruddick 1985). Haraway's socialism led her to conclude that such "feminists" were doing little more than valorizing what men had traditionally held against women without addressing the underlying source of gender inequality. These feminists' understanding of "equality" amounted to the conversion of inequality to incommensurability rather than a mandate to change the self-understandings of *both* men and women on behalf of a more comprehensive and mutually recognized sense of human equality.

However, Haraway's way of addressing this quite genuine conceptual and political problem with feminism has carried its own liabilities. In particular, she has favored an approach exploited most brilliantly and systematically by Jacques Derrida, but otherwise characteristic of French thought from the 1960s onward, which "erases" such classic binary oppositions as male/ female and machine/organism to reveal a fundamental indeterminacy of being. It is in this space of indeterminacy that a more playful attitude can be adopted toward terms whose clarity in the past served to define and regulate the human condition. In this spirit, Haraway has celebrated the "hybrid" as a being whose identity is ambivalently related to two distinct species sources. The political strategy behind this boundary blurring is that because reality does not naturally conform to regimes based on binary oppositions, the only way such regimes can be enforced is through forms of violence, both large and small, both physical and symbolic. By conducting oneself in a manner that is not easily reducible to one or the other side of the binary, the regimes can be resisted and their capacity for violence can be revealed and thereby subject to critique. At least, that was the idea.

To Haraway's 1985 audience, the radical nature of this move outside the feminist movement would have also been apparent. In particular, it refused the "positionist" politics associated with the existentialist strand of French postwar thought, epitomized by the "hard left" activism of Jean-Paul Sartre and popularized across university campuses in the Western world in the late 1960s as "If you're not part of the solution, you're part of the problem"—as powerful a statement of binary politics as any. (Full disclosure: Fuller was tarred with the brush of "positionism" early in his career: cf. Woolgar 1991.) "Gender performativity" of the sort soon to be championed by Judith Butler (more about which below) was about creating new in-between and (crucially) irreducible identities between the stereotyped male and female ones. Thus, "drag" came to be interpreted not as one sex pretending to be another but as demonstrating an identity that can thrive without being aligned to one or the other sex. This was reasonably seen as a "softening" of the left in the direction of antistatist, even libertarian politics—especially if one recalls Foucault's role as a demystifier of the state as an enforcer of binaries in the name of progressive promises. But we shall see that this is quite different in spirit from a *transgendered* identity politics, which Martine Rothblatt has artfully leveraged into a platform for establishing a transhumanist political and legal regime. Her vision effectively presupposes a substratum-neutral conception of content ("information," if you will) as the basis of identity (her word is "consciousness") that can be embodied as male, female, carbon, silicon, and so forth according to certain specified conventions.

HARAWAY'S DIFFICULTIES WITH THE PRESENT AGE: THE QUESTION OF THE TRANSHUMAN

In 2006, the British sociologist Nicholas Gane conducted a revealing interview with Donna Haraway for *Theory, Culture and Society* in the form of a belated reflection on the twentieth anniversary of her most famous work, the "Cyborg Manifesto" of 1985, where she notoriously declared that she would rather be a cyborg than a goddess (Gane and Haraway 2006). Her original intent was to celebrate the multiple prisms through which "woman" may be seen that broke free from the conventional connotations of the "male/female" binary. In her own work, Haraway appealed to primatology, cybernetics, and science fiction for examples that exploded the convention. At the time, her move was significant because it forged clear alliances between recent science and the feminist cause, which had been previously lacking or outright denied (e.g., Merchant 1980).

In addition, Haraway was ensconced in the History of Consciousness program at the University of California, Santa Cruz, where she played into an emerging "queer" consciousness. This involved the political translation of the work of the original female voices of French poststructuralism from the late 1960s—Monique Wittig, Hélène Cixous, Luce Irigaray, and Julia Kristeva—which was especially strong in the San Francisco Bay area of the 1980s. This strain of thought soon received its synthetic American expression in Judith Butler's 1990 *Gender Trouble*, a work strongly endorsed by Haraway for its clear detachment of gender identity from sexual origins. Haraway's proximity to the crucible of "queer theory" provides a background context for the easier uptake of her texts vis-à-vis those of her contemporary, Katherine Hayles, another interloper from the sciences to the humanities who canvassed many of the same themes in an equally difficult style, often with greater technical sophistication (Williams and Fuller 2014, 163–64).

However, over the past quarter century, it has become clear that Haraway understands the "detachment of gender identity from sexual origins" in a particular way—one that has made her hostile to transhumanism, which she dubs a "techno-masculinism of a self-caricaturing kind" (Gane and Haraway 2006, 146). The image of the cyborg, which Haraway had taken as mainly a metaphor for understanding gender liberation has become, as far as she is concerned, a license to fetishize mechanism. This is a charge that she lays at the doorstep of not only, say, Ray Kurzweil but also Gilles Deleuze, whose fascination with humanity's constitution by machines both large and small leads him (apparently) to mock older women and their dogs (143; referring to Deleuze and Guattari 1987, 232–309). Perhaps given her own more recent preoccupations, including the integral character of "companion species" in the human life-world, she takes this as a personal insult (Haraway 2003).

In Deleuze's defense, it is worth noting that, he was a former student of Georges Canguilhem, the man who taught history and philosophy of science to both Foucault and Deleuze at the Sorbonne. Thus, Deleuze may have been extending—in his own inimitable way—Canguilhem's (2008) identification of the mechanistic tradition from Descartes onward with the drive toward a conception of life that, *contra* Aristotle, need not appear in animal form: a *bios* that need not be *zoos*, to put the point in the terms popularized by Giorgio Agamben (1998). In this context, Cartesianism—whose empirical legacy includes experimental medicine, the brainchild of that arch vivisectionist Claude Bernard—looks "proto-transhumanist" in viewing animal life as simply a means for humans to acquire knowledge and ultimately control the fundamental (godlike) process of life itself. From that standpoint, evolutionary theory relates to biology as history relates to social science: no less, but also no more. In other words, for the die-hard Cartesian, "evolutionary theory" is a somewhat inflated name for natural history, the most interesting aspect of which is discovering the possibility space within which nature has played itself out over time. This is very much the spirit in which French molecular biologists have made Nobel Prize-winning contributions without feeling the need to genuflect to Darwin (e.g., Monod 1974).

All of this could not be further from Haraway's own repeated calls to "historicize" humanity's relationship to both biology and technology. However, "historicism" has been a fickle friend to the progressive spirit with which even the later Haraway still identifies. In particular, her combined celebration of evolutionary theory and condemnation of transhumanism seems to reflect what might be called *speciationism*, the meta-level equivalent of what Peter Singer dubbed "speciesism," the privileging of humans over other animal species. In this case, what is being privileged is the means by which one becomes a species—namely, through the evolutionary process. Thus, Haraway has no problem with—and indeed glories in—the fact that nature has always thrown up incredible diversity through the evolutionary process. However, if humans wish to pursue diversity outside the evolutionary process, say, through xenotransplantation and prosthetics, then Haraway follows the lead of past historicists in questioning the motives and consequences. In short, an acceptance of species diversity is by no means a license for species change—or, for that matter, gender change; hence, her speciationism (cf. Miah and Rich 2008, 33–35). In this respect, Haraway remains attached to the idea of the "wisdom of nature," which, to be fair, is sometimes found even in card-carrying transhumanists who worry about our intervening too radically in evolutionary processes (e.g., Bostrom and Sandberg 2009; cf. Fuller and Lipinska 2014, ch. 3).

On closer inspection, Haraway's "techno-masculinist" charge against transhumanism amounts to what Erik Davis (1998) originally identified as the movement's "techgnosis," namely, the prospect of completely replacing the biological body with a silicon-based version that retains and even amplifies one's distinctly "human" features, which are presumed to be only contingently—and perhaps only imperfectly—tied to the bodies of our birth. If the *telos* of transhumanism is understood in this way, then it is easy to see how *transgenderism* is a way station to transhumanism. If someone believes that one is a woman trapped in a man's body, then why not believe that one is a potentially infinite spirit trapped inside a finite material encasement? That, after all, was the central claim of Gnosticism, a heresy that dates from Christianity's origin. The logic is that the "morphological freedom" (to use the transhumanist jargon) that is afforded to you by the ability to switch gender identity is only the opening move to a more general ability to switch material identity altogether—in this case, from carbon to silicon.

THE ROAD FROM TRANSGENDER TO TRANSHUMAN: THE QUESTION OF WOMAN

The most articulate proponent of leveraging gender-performative identity into transhumanism has been Martine (formerly Martin) Rothblatt, who as of this writing is the highest-paid female CEO in the United States—and until 1994 was legally classed as a man (Miller 2014). The latest phase of that existential trajectory, which began with Rothblatt (1995), is Rothblatt (2014), a visionary attempt to extend the current legal regime to encompass two new sorts of entities: the *mindclone* and the *beman,* discussed below. What is most striking—and what Haraway would probably find most disturbing—about Rothblatt's characterization of these entities is they are defined in terms of criteria of recognition, regardless of how they are materially constituted. Whereas Haraway's fascination with the Turing Test—and the replicant Rachel—turns on the difficulty in distinguishing human from nonhuman, Rothblatt is more interested in the simple fact that passing the test confers ontological status. To be sure, Rothblatt would have mindclone candidates undergo a psychological gestation comparable to that undergone by transgender candidates, given the far-reaching consequences of acquiring a new ontological status (Rothblatt 2014, 102–3).

A "mindclone" is a being specifically designed to be a copy of someone else, typically oneself. What matters here is that the integrity of the thing transferred is retained in the target. This is a clearly "informational" way of thinking about the continuity of identity over time and space, which John Locke would have recognized as a criterion for "personhood." For Locke a common "memory" provided the relevant source of continuity, but nowadays what that might mean is seriously open to interpretation. But there may come a point when the mindclone has acquired sufficient experiences of its own—and those of others—that it is effectively an autonomous being and not simply an extension of the original source. In that case, its status becomes that of a "beman" (a contraction of hu*man be*ing). A test case is BINA48, a "gynoid" that Rothblatt, as postmodern Pygmalion, had constructed to look like her life partner, casually described as a "robotic torso with a software mind" (Rothblatt 2014, 203). In fact, BINA48 has been programed with not only the Rothblatt spouse's memories but also a regularly upgraded learning capacity, so that in principle it (or she) could develop its own standalone identity, as it devises novel outputs based on continual reintegration of the inputs it receives from its own unique set of interactions, typically conversations.

How, then, to characterize the difference between Haraway's cyborg feminism and the transgendered transhumanism represented by Rothblatt? The distinction between *analog* and *digital* in information theory provides a set of coordinates for charting the relevant difference in sensibility. The distinction originally referred to how ("analog") waves of electromagnetic energy are converted into a set of ("digital") numbers that convey information to humans—as in the case of wireless radio technology. Both Haraway and Rothblatt generalize this distinction, so that the analog/energy side stands for "nature," whereas the digital/information side stands for "culture." Haraway sees things from the analog side, Rothblatt from the digital side. (A profound anticipation of thinking of nature/culture in analog/digital terms is Wilden 1972, a subtle and erudite work in which nevertheless the relevant feminist strands of thought are conspicuous by their absence.)

For Haraway, gender norms invariably constrain and filter the much richer sense of reality offered by nature. Her solution is to regain an openness to that richness, as may be found in both companion animal and native American consciousness, to cite Haraway's favorite examples. In contrast, Rothblatt stresses the positive features of culture's digital character, which allows for switching between distinct embodiments, in search of a socially recognized mode of being that matches one's mental orientation. It is telling—in a way that might horrify feminists—that Rothblatt holds that each individual has a distinctive "style" of being that may, depending on circumstances, be better suited to a male or female form, as if gender as such were nothing more than "being in drag," an impression reinforced in Miller (2014), where Rothblatt admits that she nowadays finds it easier to drop the conventional female "made up" appearance, despite having undergone the corresponding sex change.

There is a world of difference between observing that the male/female binary obscures an array of in-between states (à la Haraway) and promoting an open-door policy on who or what can occupy the "male" and "female" positions (à la Rothblatt). The next logical step in Rothblatt's nominalist approach to identity would be to disavow an analog original altogether, as implied by the prospective existence of *infomorphs*, information-based organisms whose identities are formed entirely by incorporating data from those with whom they come into contact (Mason 2015). To be sure, these creatures are designed with a capacity to learn that may improve through continuous use. But unlike a mindclone, an infomorph is not programed with any specific identity, let alone that of its creator. Any identity it acquires is the exclusively the product of its interactions in cyberspace, very much in the spirit of Locke's conception of the mind at birth as a tabula rasa. It is as if these entities were made to be orphaned. Not surprisingly, significant legal issues arise in terms of how one might hold an infomorph—or, more to the point perhaps, its creator—accountable for anything it does, especially if it appears to have appropriated someone else's identity (O'Hara 2015).

DOES TRANSHUMAN/POSTHUMAN REINSCRIBE MALE/FEMALE?

It is easy to see the opposing attitudes of Haraway and Rothblatt to transhumanism as a twenty-first-century reinvention of twentieth-century gender politics, with Haraway taking the feminist and Rothblatt the masculinist stance. Moreover, when women have ventured a metaphysical vision "beyond" the human, they have preferred "posthumanism" to "transhumanism." Indeed, as of this writing, according to Google scholar, Katherine Hayles's 1999 book, *How We Became Posthuman,* is by far the most highly cited work in the posthumanist literature. (Second place belongs to the decidedly antiposthumanist work, Francis Fukuyama's *Our Posthuman Future.*) The most obvious reason for the "feminization" of posthumanism is its explicit decentering of the "human" as the locus of value, where "human" implies the Enlightenment vision of the dominating male subject that culminates in the Nietzschean "superman," something openly embraced by transhumanists. In other words, the posthumanist takes "human" to be so tainted with masculinity that true gender justice can be achieved only by removing the species privilege of *Homo sapiens* altogether. Thus, the various ways in which the sexual division of labor is played out across the rest of the animal kingdom become grist for the posthumanist's mill.

To test this hypothesis, Veronika Lipinska, coauthor of Fuller and Lipinska (2014), interviewed two prominent women who might be reasonably called "transhumanist": (1) Natasha Vita-More, author of the original "Transhuman Manifesto" of 1983, which introduced the concept of "morphological freedom" as a horizon opened to humans by virtue of advances in modern technology, and long-term life partner of Max More of "proactionary principle" fame (Lipinska and Vita-More 2014). (2) Rachel Armstrong, medical doctor, science fiction author, and nowadays champion of "living architecture," the idea that humans can build more ecologically secure environments by exploiting what nature "naturally" does (Lipinska and Armstrong 2014). Of the two sets of responses to five identically posed questions about how they relate themselves to feminism and transhumanism, Vita-More's were more like Rothblatt's and Armstrong's like Haraway's. We conclude by highlighting some of the salient differences. (See Appendix for the five questions.)

While neither Vita-More nor Armstrong denies the value of humanism, Vita-More explicitly regards transhumanism as an extension of humanism. Even her conception of the "posthuman" has a humanist spin, namely, "potential futures for the human when it is no longer exclusively a biological animal." For Vita-More, ecological matters arise only in context of "existential risks" posed by, say, global warming—but without positive reference to nature or other creatures. This is in striking contrast to Armstrong, who sharply distinguishes transhumanism and posthumanism, identifying herself with the latter, which is focused on "ecological being," whereby "humans are not 'apart' from an ecological existence but woven into its fabric and deeply entangled within it." This difference in perspective was also in evidence when it came to the two women's relationship to feminism. Whereas Vita-More defined her brand of feminism in terms of "gender selectivity—that gender is a state of mind, largely based on personal preferences and social and cultural differences, rather than on an individual's genitalia," Armstrong gave a more qualified response, saying that "while [her politics] are obviously placed within a history of feminism, they are more interested in diversifying and disseminating forms of power within society rather than polarizing debate around established dichotomies." On the evidence of these two prominent female thinkers "beyond the human," it would seem that the trans/posthuman distinction is well on its way to reinventing the differences in worldview that has characterized the male/female divide.

APPENDIX

The five questions asked by Veronika Lipinska to Natasha Vita-More and Rachel Armstrong in November 2014:

1. As a speaker and public intellectual, you deliver the transhumanist message worldwide. Do you, in your opinion, make gender points in transhumanist settings? If so, what gender points do you make?
2. Why did you decide to be a transhumanist? Why not just a humanist or a posthumanist?
3. Do you identify yourself as a feminist? Is gender identity something important to you?
4. When online, for example, on chats and forums, do you admit your gender identity or do you keep it concealed? Why?
5. What are the most important topics that you discuss with other transhumanists? Do you feel that there are topics you would like to discuss but cannot?

BIBLIOGRAPHY

Agamben, Giorgio. *Homo Sacer: Sovereign Power and Bare Life*. Palo Alto: Stanford University Press, 1998.

Bostrom, Nick and Anders Sandberg. "The Wisdom of Nature: An Evolutionary Heuristic for Human Enhancement." In *Human Enhancement*, edited by Julian Savulescu and Nick Bostrom, 375–416. Oxford: Oxford University Press, 2009.

Canguilhem, Georges. *Knowledge of Life*. New York: Fordham University Press, [1965] 2008.

Daly, Mary. *Gyn/Ecology: The Meta-ethics of Radical Feminism*. Boston, MA: Beacon Press, 1978.

Davis, Erik. *TechGnosis: Myth, Magic and Mysticism in the Age of Information*. San Francisco: Harmony Books, 1998.

Deleuze, Gilles and Félix Guattari. *A Thousand Plateaus*. Minneapolis: University of Minnesota Press, [1980] 1987.

Fuller, Steve and Veronika Lipinska. *The Proactionary Imperative: A Foundation for Transhumanism*. London: Palgrave Macmillan, 2014.

Gane, Nicholas and Donna Haraway. "When We Have Never Been Human, What is to be Done?" (Interview with Donna Haraway.) *Theory, Culture and Society* 23, nos. 7–8 (2006): 135–58.

Genova, Judith. "Turing's Sexual Guessing-Game." *Social Epistemology* 8, no. 4 (1994): 313–26.

Haraway, Donna. "A Cyborg Manifesto." In *Simians, Cyborgs and Women*, edited by Donna Haraway, 149–81. London: Routledge, [1985] 1991.

Haraway, Donna. *Companion Species Manifesto*. Chicago: University of Chicago Press, 2003.

Lipinska, Veronika and Rachel Armstrong. "Interview with Rachel Armstrong." Conducted on-line 11 November 2014.

Lipinska, Veronika and Natasha Vita-More. "Interview with Natasha Vita-More." Conducted on-line 15 November 2014.

Mason, Luke Robert. "Data Made Flesh: Artificial Persons with Artificial Organs." London: Royal Academy of Arts, 2015.

Merchant, Carolyn. *The Death of Nature: Women, Ecology and the Scientific Revolution*. New York: Harper and Row, 1980.

Miah, Andy and Emma Rich. *The Medicalization of Cyberspace*. London: Routledge, 2008.

Miller, Lisa. "Trans-Everything CEO." *New York Magazine*, 7 September 2014.

Monod, Jacques. *Chance and Necessity*. Translated by Austyn Wainhouse. London: Fontana, 1974.

O'Hara, Dan. "Internet journalism and the rise of a new satire." *Open Democracy*, 13 April 2015. www.opendemocracy.net/ourkingdom/dan-o%27hara/internet-journalism-and-rise-of-new-satire.

Rothblatt, Martine. *The Apartheid of Sex*. New York: Crown Publishers, 1995.

Rothblatt, Martine. *Virtually Human*. New York: St. Martin's Press, 2014.

Ruddick, Sara. *Maternal Thinking: Towards a Politics of Peace*. Boston, MA: Beacon Press, 1985.

Wilden, Anthony. *System and Structure*. London: Tavistock Publications, 1972.

Williams, Rhys and Steve Fuller. "Humanity 2.0." (Interview with Steve Fuller.) *Paradoxa* 26 (2014): 159–72.

Woolgar, Steve. "The Very Idea of a Social Epistemology." *Inquiry* 34 (1991): 377–89.

Chapter 24

Beyond Black and Green

Children Visioneering the Future

Emilie Whitaker

Steve Fuller (2013) has written extensively in recent times on what he foresees to be "ninety degree revolution" in politics away from the traditional left/right distinction toward a politics where proponents divide along heavenly or earthly orientations. Here, he describes "upwinging Blacks" and "downwinging Greens." Such binaries are presented as schemata, yet the implication for Fuller's own theologically imbued epistemology is clear. The Blacks are the future with their proactionary, risk taking, embracing of science and technology, pushing humanity out of a stasis. The Greens are dismissed, in the main, as earthly luddites. Yet, Fuller (2013) presumes not only the evident status of these binaries, but states on this political shift, "So far I have portrayed this ideological rotation from the standpoint of a younger generation that accepts it as a given." The question at hand for this chapter is simply, is this the case? Do young people align more or less with these orientations and the presumed adherence to proactionary or precautionary approaches particularly with regard to science and technology? How do they "visioneer" their futures? Visioneering is where young people both describe their vision of the future and posit the "engineer" aspects of the neologism, to answer how we got there (McCray 2012).

METHODOLOGY

To investigate the visioneering activity of young people, the author and Steve Fuller have undertaken exploratory ethnographic and inductive work with groups of young people. The pattern of this burgeoning study has been to undertake semistructured group interviews with groups of children aged twelve to sixteen as part of existing events themed around science and technology. These group conversations utilize the settings, the props, and the talks already given to open up discussion on what the future may hold and how they come to understand it in this way. In addition, to informal, digitally recorded interviews, we devised visual methods creating "postcards from the future wall" where participants describe a potential future and send it back to the present. The study is exploratory in scope and takes its ethical orientation from the new sociology of childhood that places emphasis on the participatory and agential focus of work with children and young people (Jenks 2004; James and Prout 1990). The interpretive lens adopted takes children and young people actors in their own right, as meaning makers, albeit enmeshed in temporal conditions not of their own making.

The small snapshot of data provided here is taken from an event whereby over seventy-five children interacted with us either in group conversations or through utilizing the "postcards from the future" wall. We have accumulated over six hours of group data from the event, so all that can be presented here are the very initial themes that punctuated the day. Interestingly, in Fuller's schemata themes were predominantly "green" rather than "black." In the main, groups were concerned with near-future visioneering within their lifetime and placed emphasis on the Anthropocene over the heavens.

Two group discussions are presented here as crudely encompassing what at first glance appear to fall neatly into the "upwinger"/"downwinger," Black/Green classifications charted by Steve Fuller (2013) in a widely read piece for *Aeon* magazine. Both discussions focused on the kinds of existential threats that form the backbone of dystopian futures—one on the impacts of environmental disaster wrought by climate change, the other on the potential impacts of a range of common transhuman proclivities—augmentation, genetic engineering, and artificial intelligence. But they complicated Fuller's undulating poles. In his original formulation he implies not only an earthly or celestial/trans orientation, but links such orientations to an embrace of or resistance to technologies. Thus, downwingers are cast somewhat disparagingly as resistant to transcendence of many kinds—preferring instead to cling to some Aristotelian essentialized human nature, the mark of which is our biological encasing on earth. They embody the precautionary principle enshrined in legislation following the atrocities committed in the twentieth century.

In contrast, upwingers receive much more favorable treatment as the bringers of Humanity 2.0, creatively destroying old mores to pull us out of our current ideological stasis. Upwingers are those who enact the proactionary principle (Vita-More 2013; Fuller and Lipinska 2014); they are "proactionaries" taking calculated risks as part of a program for human progress, where the capacity for progress is taken to define us as a species.

TECHNOLOGY TO REMAKE, REMOLD, AND REGROW THE NATURAL

> Why do nature and technology have to be in competition, combat? Why don't we use what we know about biology to rebuild nature? I mean we can grow skin, map and alter our genetic code, grow modified crops, why not turn that to support the natural. It's all living.
>
> Yeah. Why is it that environmental concerns are ignored or put back all the time? I mean it's pretty arrogant to think we can just up and leave on some imaginary spaceship, we are here and now.

These quotes were taken from a group discussion with three fifteen-year-old girls. Two wanted to go into medicine and one was undecided although she was interested in the interplay between "biology and engineering." The three young women were all concerned about climate change and the impact of this not only on our natural environments but also on our geopolitical engagements with each other. One of the group's girls told us,

> It's not just about companies and coal, it's about what the effect is of disasters in one part of the world on another. So I can see real issues with . . . like . . . environmental refugees because their villages have flooded or a tsunami has knocked out their industry.

In the visioneering work undertaken by young people on the day, such concerns about climate change, environmental crisis, and poverty were not marked by a technological resistance. There was no talk of a quest for holism that can only be realized in "returning to nature" and rejecting the malevolence that science has wrought. Indeed, given the patriarchal narrative that has historically aligned women with nature to serve a politics of oppression, this is an important point to make. Plumwood points out (1997, 19), "Feminine 'closeness to nature' has hardly been a compliment."

In fact in the accounts given by this group, these "downwingers" were comfortable in harnessing the power of technology to protect the earth, to almost instigate a second natural flourishing, not dissimilar to some of the ambitions espoused by "Living Architect" Rachel Armstrong (2010). When asked about the potential for synthetic biology one of the young women responded,

> You know, why not harness all that we know about biology and synthetic environments and use that to repair the planet? I just don't see why there's "man made" in one corner and "the natural" in the other. And we just abandon nature.

Their accounts shared commonalities with the rise of ecomodernity as an alternative to environmentalism. The ways in which the young women emphasized the human consequences of environmental degradation point not to an abandonment of science but rather to its interspersion with reflexive modernization. That is under ecomodernism, "science should be demonopolised and democratised and redirected toward a social rationality" (Bäckstrand 2004, 700). Befitting such a humanity-oriented greenness, they were not afraid to make a political case for living differently; this was a rarity as the data collated so far indicated an absence in the role of the state in these future dystopias:

> You've got the government to do more, I mean, people keep saying it's going to take a long time but if you don't start you never finish. If you start now we'll be done in 15–20 years. We need scientists to get into politics, they have the knowledge, if they go into politics they can spread that knowledge. They need to stop being scared, their insecurity is less important than what's going on in the world. They need to stop thinking of themselves and think about all the people that would benefit. People are too selfish to think about the bigger problems.

Those young people who were passionately concerned about climate change described a human-centered rather than geo-centered world; there was no mention of animal sentience and the bestowing of rights to nonhuman creatures. In the group discussion where downwinging played a central role, the young women making the case for climate interventionism seemed to be making it on a vitalistic rather than Darwinian premise. Such a vitalism may form a better entry point into the cartography being mapped by these young people, as their lives are complexly mediated by the blurring of the body with technology, the ecological with the manufactured. Their sophistication seemed to contest the Green/Black binary by refusing to oppose nature to culture, environment to society, and art to science. The creative approach to sustainability, of utilizing technology to remake the natural undermines the sharp distinction between "black and green" as somehow indicative of a proactionary or precautionary stance to science and technology.

EMBODYING RISK: ANTIFRAGILITY AND RESILIENCE THROUGH GAMING

> *Steve Fuller:* So how do you see the future?
>
> Very dark really, things going down. Underdeveloped countries, pollution, resource crises . . .
>
> *Steve Fuller:* You sound pessimistic about the future!
>
> Yeah I think there are a lot of big problems.

Such dystopian visions emerged most readily in references to gaming, which emerged as a dominant theme in this group of young men. Notable mention went to the cyberpunk-styled game *Deus Ex: Human Revolution* set in a near-dystopian future. *Human Revolution* tackles transhumanist themes through the eyes of the protagonist, an employee of a biochemical human augmentation firm, as he considers whether humanity's reach has exceeded its grasp. Its societal setting is cataclysmic: corporations have greater power than states, corruption is rife, and rebellion put down with brutal violence.

> I play a game set in the future, *Deus Ex: Human Revolution.* There's global terrorism and someone releases a virus. People are getting synthetic arms, nanotechnology all kind of augmentation. It'll happen, you know. I've seen these things in the news already, well something like that. The thing is, if we do get these augmentations then it's going to be a taboo to be normal. That is a realistic possibility. Also, in the game you need money or access to resources, there's no NHS you know. So only rich people benefit.

Another game mentioned was the post-apocalyptic game *Fall Out*:

> *I think a possible future is like Fall Out because of wars and resource shortages.*
>
> *Steve Fuller:* Do you really think nuclear war is a possibility?
>
> Definitely. Nuclear war is a possibility for the future; it'll just be started over different issues than before. Like Russia invaded Georgia and no one cared, militias are growing in Crimea. In the future these little skirmishes become more important as resources shrink.

These young men spoke of the gaming experience as a tool to furnish their visioneering activity alongside their interest in the practice of formal scientific inquiry and their own personal hopes and ambitions. The gaming activity offered up a language and a set of tests—to consider difficult "what if" scenarios. It was as though the practice of gaming enabled a relationship with risk and the ethics of risk to be contemplated and explored. It offered a visceral window into visioneering practice as gaming was described as something experienced and embodied not merely thought or seen.

Qvortrup (2003) uses the concept "hypercomplex society" to describe how digital 2.0 communication technologies enmesh the local intimately with the global. In this hypercomplex society culture becomes, according to Poster (2003), a heady mix of multiple meanings that young people seem to negotiate, from the early analysis of our findings, with a great deal of ease. The interaction of mobility with information, social movements with geographic space through the politics of the hashtag proffers a very intimate yet global hypercomplex society. The four fourteen-year-old boys considered the consequences of flooding in Bangladesh, the fallout of genetic experimentation, and the ethics of epidemic through their avatared prism of gaming through near-future contexts provided by *Deus Ex: Human Revolution*. Global themes are experienced locally. While such visions have been presented as evidence of a dystopian generation, weary already with the atrocities and unfairness of the world, perhaps these themes are present in accounts because the world reaches into and mediates so many spaces experienced by young people. Moreover, they reach back. From smartphones to Second Life, Twitter to rolling 24-hour news, there is a degree of "hyperreality" (Baudrillard 1983) in the visioneering accounts presented to us. The four young men in this group all owned smartphones and tablets, two had Twitter accounts, and all were using online apps and platforms to interact with friends and share information across time and space. One told me, "I could not live without it [smartphone]. So much of my life is on it and through it." In the case of many participants engaged with in this session, information and communication technologies have become "arenas for social experience" (Stone 1995, 15). In this regard, the Baudrillardian hyperreality (1983, 11) of distractive symbols and codes also holds the potential for extension of the self into the world. This interpellation may enable young people to equip themselves with cognitive and experiential tools to understand and experience risky visions under terms of relative safety.

It is in this interactive reaching in and out that the duality of the "real" and "virtual" begins to disintegrate, as Lévy puts it (1998, 23): "Consider the simple and misleading opposition between the real and the virtual." The dystopian experiences navigated through gaming held corporeal significance within the hyperreality, one of the participants simply stated,

> You feel it as you get used to being him [Adam Jenson—protagonist]. The noise of it, it like speeds your heart rate up, you know something's going to happen and you know you need to respond. Fight or flight isn't it?

Thus, it would be erroneous to posit a neat distinction between an embodied "present" body and a disembodied, gaming body. Rather, gaming offers scope to experiment with the limits and dimensions of the self, including material and corporeal sensations.

Increasingly, sophisticated interactive gaming cultures enable transgression and transmogrification of the self and body through visioneering settings and experiences. Risk can be felt, experienced, mitigated, and accelerated, opening up new worlds of cognitive consideration and vision. The expansion of experience encountered through digital worlds is supportive of Ong's (1982) argument that the appropriation of new forms of expression alters the very horizons for human thought and cognition. Here, the boys' use of gaming enables them to work through, in a pseudoembodied sense, alternate conceptualization of the future and their attitudes toward it. In this regard, perhaps Fuller and Lipinska's (2014) call for a "proactionary" imperative for public policy could find utility in gaming cultures as sites for trial and error, risk taking, and risk making. Such a program would take the idea of the "cyborg citizen" quite seriously (Gray 2001) as young people are encouraged to take risks as part of a dual program of "antifragility" and inoculation. With regard to children and young people in particular, we can only understand risk in relation to resilience (Ungar 2011; Daniel 2003). Bolstering resilience through exposure to managed risk has been put forward as a method for supporting the adaptive quality of resilience in young people (Empson and Nabuzoka, 2004). Addressing resilience in young people Daniel (2003, 7) describes, "Resilience is not simply an absence of psychological symptoms despite having experienced adversity, it is the possession of a positive adaptive ability that enables a person to feel competent despite risky living conditions." Masten et al. (1990, 426) defined resilience in children as "the process of, capacity for, or outcome of successful adaptation despite challenging or threatening circumstances." Thus, it is not the absence of risk that undermines resilience, but the managed exposure to and successful negotiation of the encounter. Here Taleb's (2012) concept of "antifragility" may help us to understand the utility of visioneering activity through gaming culture. In Taleb's conceptualization, the "antifragile" agent does not merely withstand challenge as it arises; she seeks to improve her current condition as the environment changes, without clinging to any preordained sense of normality. The antifragile agent engages in both spread betting—ensuring multiple options are covered—and visioneering, exploring the action and consequences of routes taken and not taken. The inoculation approach encompassed in idea of "antifragility" (Taleb 2012) captures the active conceptualization of resilience as adaptive and learnable quality.

In addressing Fuller and Lipinska's (2014) demand for a policy program supportive of risk taking in a risk-anxious culture epitomized by fears about and for young people, digital spaces may provide a test case. In the sense of a program for proaction, gaming enables a vicarious yet embodied engagement with the experience of risk taking without the fear of existential precarity. Avatars become less a representation, a symbolic extension of the digital self and more a site for potential embodiment, particularly as gaming culture and technology advances to become still more corporeal.

CONCLUSION

The visioneering work of these young people rejects the kind of political generational talk as filibuster arguing instead that "we need to think differently in order to live differently" and "more scientists need to get into politics."

Their accounts challenge common kinds of epistemological boundary work, the policing of the possible from the impossible, by collapsing the now with the almost-now, the existent with the becoming. In this regard, whether exploring the scope of synthetic biology to literally rebuild nature, or considering geopolitical maneuvers in cyberspace, the accounts of young people indicate a sympathy with Haraway's "natureculture" (2003). In this vein, young people as digital natives and cyborg citizens in-the-making denote Turkle's (1995, 21) "transgressive mixture of biology, technology, and code." In talking about their experiences in the social realm of gaming they consider alternative dystopian futures and confront technological advances within an ethical and social framework. Some of the mechanics and poetics of this visioneering are also akin to Kelly's (1979) "double-edged vision." This concept speaks of the power of hybridizing lucid argument with political and personal passion leading to the creation of alternative social blueprints. In taking an agential focus to the visioneering work of the young participants, the burgeoning data suggests the importance of epistemological humility in the subtleties of visioneered potentialities, subtleties that challenge our neat constructions of Black and Green, Up and Down, Trans and Posthumanity.

BIBLIOGRAPHY

Armstrong, Rachel. "Living Architecture." *News in Conservation*, International Institute for Conservation, 20 (2010): 4–5.

Bäckstrand, Karin. "Scientisation vs. Civic Expertise in Environmental Governance: Eco-Feminist, Eco-Modern and Post-Modern Responses." *Environmental Politics* 13, no. 4 (2004): 695–714.

Baudrillard, Jean. *Simulations*. New York: Semiotext[e], 1983.

Daniel, Brigid. "The Value of Resilience as a Concept for Practice in Residential Settings." *Scottish Journal of Residential Care* 2, no. 1 (2003): 6–15.

Empson, Janet M. and Dabie Nabuzoka. *Atypical Child Development in Context.* London: Palgrave Macmillan, 2004.

Fuller, Steve. "Ninety-Degree Revolution Right and Left are Fading Away. The Real Question in Politics Will Be: Do You Look to the Earth or Aspire to the Skies?" *Aeon Magazine*, October 24, 2013.

Fuller, Steve and Veronika Lipinska. *The Proactionary Imperative: A Foundation for Transhumanism.* London: Palgrave Macmillan, 2014.

Gray, Chris. H. *Cyborg Citizen—Politics in the Posthuman Age.* London: Routledge, 2001.

Haraway, Donna. *The Companion Species Manifesto.* Chicago: Prickly Paradigm Press, 2003.

James, Allison and Alan Prout, eds. *Constructing and Reconstructing Childhood.* London: Falmer, 1990.

Jenks, Chris. "Constructing Childhood Sociologically." In *An Introduction to Childhood Studies*, edited by Mary Jane Kehily, 77–95. London: Open University Press, 2004.

Kelly, Joan. "The Double-Edged Vision of Feminist Theory." *Feminist Studies* 5, no. 1 (1979): 216–77.

Lévy, Pierre. *Becoming Virtual: Reality in the Digital Age.* New York: Plenum Trade, 1998.

McCray, Patrick W. *The Visioneers: How a Group of Elite Scientists Pursued Space Colonies, Nanotechnologies, and a Limitless Future.* Princeton: Princeton University Press, 2012.

Masten, Ann S., Karin M. Best, and Norman Garmezy. "Resilience and Development: Contributions From the Study of Children who Overcome Adversity." *Development and Psychopathology* 2, no. 4 (1990): 425–44.

Ong, Walter J. *Orality and Literacy: The Technologizing of the Word.* London: Routledge, 1982.

Plumwood, Val. *Feminism and the Mastery of Nature.* London: Routledge, 1997.

Poster, Mark. "Perfect Transmissions: Evil Bert Laden." *Television and New Media* 4, no. 3 (2003): 283–95.

Qvortrup, Lars. *The Hypercomplex Society.* London: Peter Lang Publishing, 2003.

Sagy, Shifra and Naomi Dotan. "Coping Resources of Maltreated Children in the Family: A Salutogenic Approach." *Child Abuse and Neglect* 25, no. 11 (2001): 1463–80.

Stone, Rosanne A. *The War of Desire and Technology at the Close of the Mechanical Age.* Cambridge, MA: MIT Press, 1995.

Taleb, Nassim N. *Antifragile: How to Live in a World We Don't Understand.* London: Allen Lane, 2012.

Turkle, Sherry. *Life on the Screen: Identity in the Age of the Internet.* London: Weidenfeld & Nicolson, 1995.

Ungar, Michael. "The Social Ecology of Resilience: Addressing Contextual and Cultural Ambiguity of a Nascent Construct." *American Journal of Orthopsychiatry* 81, no. 1 (2011): 1–17.

Vita-More, Natasha. "Contested Culture: The Plausibility of Transhumanism." *Existenz* 8, no. 2 (2013): 19–25.

Chapter 25

Prolegomena for a Theory of Justice for a Proactionary Age

Steve Fuller

Justice has been always about modes of interconnectivity and mutual accountability. The original theory of justice was based on retribution an "eye for an eye." It recalled an age when kinship was the basis of human society. In the modern era, courtesy of the nation-state, bonds have been forged in terms of common laws, common language, common education, common roads, and so forth—all of which have run interference on traditional kinship bonds. The Internet, understood as a global information and communication infrastructure, is both enhancing and replacing these bonds, resulting in new senses of what counts as "mine," "yours," "theirs," and "ours"—the building blocks of a just society. This emerging sensibility compels a reexamination of the terms in which justice is defined and delivered.

That scourge of Silicon Valley, Evgeny Morozov (2013) is undoubtedly correct when he says that the Internet is not sufficient to solve the problem of justice, today or tomorrow. But he is wrong to suggest that it is not necessary. Indeed, we need to reinvent digital versions of the bonds that have been forged over the past 250 years of nation building. Morozov's quite justifiable fear is that some of the most crucial bonds may be lost in translation. As Morozov (2011) himself first observed vis-à-vis the so-called Arab Spring of 2010, the networks forged by social media provide a flimsy basis for lasting institutional reform. Indeed, in many respects, the Arab Spring's aftermath resembles the often reactionary turn that European politics took following the abortive "liberal" print-fueled revolutions of 1848 (Mason 2012). This in turn led Karl Marx to shift his focus from the bourgeoisie to the proletariat as the agents of world-historic justice (Fuller 2012; cf. Kurzman 2008). In what follows, it may seem that I am presenting Silicon Valley as the latest such agent. Without wishing to argue that point, nevertheless I propose that emergent tendencies from that digital mecca provide a fruitful context for conceptualizing justice in our time.

IS THE MODERN PROBLEM OF JUSTICE AN ARTIFACT OF THE INDUSTRIAL AGE?

In the second half of the twentieth century, liberal political theorists often spoke of a trade-off between "efficiency" and "equity" in the administration of justice: the most efficiently run society might end up treating individuals unequally in the name of some higher sense of welfare, if others are better able to serve one's needs better than oneself—and without net loss to their own welfare (Barry 1965). This was normally understood as an argument for the state to directly benefit the rich in order to indirectly benefit the poor, as in a tax system that encourages capital investment to enable job creation. Economists call this win-win situation "Pareto optimality," after the great Italian political economist and sociologist Vilfredo Pareto. To be sure, this state need not reduce economic inequality—indeed, it may even increase it. But it would be difficult to call such a state "exploitative" because the rich can get richer only if the poor also get richer, even if not at the same rate.

While Pareto tended to be read in the twentieth century as a response to Marx, in fact his original target was John Stuart Mill's intuitive appeal, via the principle of diminishing marginal utility, that any relief of poverty requires an ethically tolerable redistribution of income. Concretely put, Mill reasoned: £1,000 shifted from a millionaire to a pauper will benefit the pauper much more than it will hurt the millionaire, given the difference in their starting level of wealth. However, the intuitiveness of Mill's argument is compromised by his assumption of a steady-state (i.e., not growing) economy and, as cognitive psychologists have more recently shown, the greater harm that is felt from regretting a loss than never having achieved a gain, even though the net result is the same. Aesop's fable about "sour grapes" captures our psychic immunity to the latter state of being (Elster 1983).

John Rawls (1971) famously tried to honor both Mill and Pareto with his "difference principle" as the main constraint on his theory of justice. Rawls argued that a just society should be organized according to egalitarian principles unless it is likely that some people will have their needs better met by others controlling a larger share of society's resources. In other words, Rawls did not merely insist that the poor end up marginally better at the hands of the rich but that their basic needs are actually met. Thus, the principle invites suspicion of any easy "trickle-down economics" that presumes that the unfettered rich will necessarily raise the standing of the poor, as Pareto has been often interpreted.

However, the common problem faced by Mill, Pareto—and even Rawls—in today's world is that none of them placed sufficient trust in real people to participate in decisions concerning their own welfare. All of the conclusions drawn—however beneficial to the people involved—are reached by largely a priori means. Whereas Mill and Pareto were quite clearly addressing a utilitarian legislator, Rawls derived his principles of justice from a thought experiment—the "original position"—that purported to have people choose the kind of society in which they would live. However, these people do not make a free choice under normal epistemic conditions. Rather, under Rawls's "veil of ignorance," they would have perfect knowledge about the general state of each possible society combined with complete uncertainty about one's own place in that society. As many liberals and libertarians immediately remarked, little wonder that people would end up plumping for a society that minimizes the worst life chances (e.g., Hare 1973). But it is not clear why this should be the decision makers' "original position," other than that it puts one in the frame of mind to reach Rawls's desired conclusions. (Fuller [1985] considers the sensitivity of our intuitions about justice to such initial epistemic conditions.)

HOW THE INFORMATION AGE HAS CHANGED THE CONDITIONS OF JUSTICE

A sense of the historical long-view helps to explain why the above line of thinking is now being displaced. The efficiency-equity trade-off underwriting late twentieth-century theories of justice presupposed an industrial mode of production, in which welfare policy was set by market- or state-based schemes of resource allocation, where the goods on offer were subject to considerable variability and discretion. The final remnant of the previous agricultural mode of production in these accounts was a sense of the economy's overall finitude, a theme picked up by later ecological economists, sometimes inspired by Karl Polanyi (1944). This "limits to growth" view of industrialism was clearly still present in Mill but already absent in Pareto, who perhaps in spite of himself anticipated the economic consensus on both sides of the Iron Curtain during the Cold War was that the best way to eliminate poverty (both at home and abroad) is to grow the economy—indefinitely. In this context, a Rawls-style wealth redistribution scheme would appear relatively painless, as everyone would be effectively getting richer through paid employment.

However, for better or worse, the "limits to growth" agenda is back on the table in the post-Cold War era. Nevertheless, the shift from an industrial to what is often called an "informational" mode of production has meant that potential scarcities are increasingly met by *substitution strategies*. These involve replacing the first-order want for a good with a second-order need that, at least in principle, may be satisfied more easily. Such strategies make sense if the cost of producing the requisite range of goods is less than continuing with business as usual—and perhaps even less than the cost of fathoming the consumer's needs in the first place, especially if depth-psychological methods are used. The relative growth of marketing vis-à-vis manufacturing in the twentieth-century corporate balance sheet testifies to this trajectory, paving the way for today's information economy (Stinchcombe 1990, ch. 5). Moreover, the trajectory is epistemologically rooted in the idea that whatever we may currently take to be true (cf. good) is ultimately a proxy for a range of values that may satisfy the "truth function" (cf. want/need) under a variety of conditions. This idea, traceable to Galileo's conception of physical laws as mathematical functions, came to be generalized in twentieth-century analytic philosophy in Quine's (1953) slogan, "To be is to be the value of a variable." (The strands from physics, logic, and economics alluded to here have yet to be joined up, but Cassirer [1923] gets you out of the starting gate.)

To be sure, in its original twentieth-century incarnation, marketing's dominant philosophy drove firms to build "planned obsolescence" into products as a means of forestalling market saturation. In practice, this resulted in unprecedented material wastage, which arguably helped to precipitate the ongoing ecological crisis that began in the late 1960s. Thus, by the late 1950s people came to believe that they needed each year to trade in their new car for a marginally better one (Galbraith 1958). This tendency continued largely unabated in the following generation with the advent of personal computers (smart devices), machines that were promoted as "vehicles" or "platforms" (i.e., vehicle-ready), not least by Al Gore's "information superhighway" slogan and the downstream idiom, "surfing the web."

At the same time, however, a counter-trend was already forming in the late 1960s, which has come into its own in our own day: It is what Abraham Maslow at the end of his life called "Theory Z," which is very much in evidence whenever people decide to buy A instead of the virtually identical B due to something they know about how A was produced or where profits from its sales will go (Maslow 1971, ch. 22). "Fair trade" and "Green" consumerism are latter-day cases in point. What is especially striking about this development is its resource-use efficiency. The marketing division no longer sees its main task as coming up with more attractive products but with more attractive *accounts of products*, as consumers are seen as mainly in the business of furnishing their inner lives. In principle, at least, the application of Theory Z implies a more ecologically sustainable version of capitalist wealth production, as the system is oriented toward producing new attitudes rather than new things. People are effectively being persuaded to pay for things above their market value simply because they carry symbolic value beyond their use value. What matters is not possession of the thing but possession of the quality that the thing represents, which is the stock-in-trade of advertising, *pace* Quine, the manipulation of the variables that define who we are.

Corresponding to this shift to "transcendent goods," as Maslow dubbed them, is a shift in intuitions about the basis for rights, which in turn has consequences for our intuitions of justice. It has been long known that Internet-based search and social media companies such as Google and Facebook manage to keep their services free at the point of delivery because they sell the data that their users routinely provide to agencies that specialize in the analysis of intelligence, ranging from corporate marketing divisions to the U.S. National Security Agency, along with experimental social scientists who may use the data to manipulate user behavior in real time. The emerging object of study is now popularly called "big data"—and its science "network science" (Pentland 2008). This is the matrix in which transcendent goods are nowadays conceived. As some recent high-profile court cases revealed, Google and Facebook users contract to have their data retained indefinitely and deployed for purposes that these companies see fit, on condition that the users themselves are not harmed. The cases turn on whether the users realized what they signed up for—and, more importantly, whether it is fair that they must do so to avail themselves of Google's or Facebook's services.

Many would wish to place severe restrictions on the access rights that Google and Facebook may demand of their users. These critics effectively extend certain intuitions about the territorial integrity of the physical self into the digital realm—call it "digital dignity," if you will. These intuitions underwrite the liberal notion of rights as an expression of self-possession. Modern law's conception of the mind's external realizations as "intellectual property" flow from these intuitions as well. Thus, in patent law, the privacy normally allowed to one's mind for incubating an idea is extended to cover a period when the resulting invention is allowed to find its place in the market without interference. That one speaks of "interference" in this context implies that if others were involved in developing the invention, the inventor himself or herself might lose some of his or her capacity to do so. In other words, the sphere of exploitation for an invention is presumed to be finite and hence given to rivalry. It is just this exclusivity of entitlement that renders products of the mind *private* property. However, it remains an open question whether "privacy" in this sense actually promotes people's productive capacity. At most, it temporarily shields their own incapacities, given that patents eventually expire. But would it not make more sense to organize the regulation of intellectual property around the fact that inventors may not be best placed—either socially or cognitively—to elicit the full potential of their inventions, but at the same time they should not be unduly disadvantaged by that fact?

A NEW SENSE OF JUSTICE FOR A NEW TIME?

At this point, let us return to first principles. Liberalism classically upholds the ideal that each person should be free to the same extent as others, and the just society is one that allows this state of affairs to be realized. The state is thus licensed to maintain the conditions for what we nowadays call "equal opportunity." But people should not lose this opportunity altogether simply because they have made a bad decision or have suffered the consequences of actions by others. Here the law can be freedom's friend through a shift in the basis of rights—the legal mainstay of liberalism—from *property* to *liability*. The *locus classicus* in legal scholarship is Calabresi and Melamed (1972), which has been recently developed to promote a "proactionary" social order that is open even to radical self- and social experimentation (Fuller and Lipinska 2014, ch. 4).

The shift turns on a question: *Where to draw the line between the legitimate exercise of my own rights and my preventing you from doing likewise?* Rights are normally understood in terms of property relations, extending to "intellectual property." Indeed, the idea of "invasion of privacy" suggests a mind ought to be treated like a piece of real estate, over which others may not pass unless explicitly authorized. But real minds are of course porous entities that cannot develop their capacities without the outside world regularly impinging on them, a concept increasingly called the "extended mind" (Clark 2008). Perhaps, then, rights should be assigned differently—namely, in terms of whether a mind gets what it needs from those who impinge on it to continue developing its own capacities. Once you start thinking this way, you're on your way to making the relevant legal shift toward a view of rights based on *liability*.

In certain areas of the law, the shift is regularly observed. For example, copyright law entitles a radio station to play my song without seeking my permission each time because it has already made a general agreement to pay song writers a fixed amount with each play. In this way, the station acknowledges its debt to song writers in a way that also facilitates their producing music in the future. The same principle may be extended to information gathering of the sort pioneered by market researchers, perfected by intelligence agencies, and nowadays increasingly gathered by digital means (Lessig 2001, ch. 7). I would then not worry so much about whether you know something about me that I thought only I knew, or perhaps even did not know about myself. But I would worry more about whether I am materially disadvantaged by your having access to such knowledge about me. This shift in perspective would in turn entail formal negotiations over how—and how much—remuneration or compensation is entitled to those mined for information. Perhaps these negotiations would be handled by collective representatives, as in the case of musicians' unions. As it stands, such matters are handled much too informally to be reliable. Thus, Google simply presumes that your free use of their search engine entitles them to mine and utilize any and all data relating to your usage. The National Security Agency similarly presumes that monitoring the phone call patterns of ordinary citizens is a fair price for our living in a better protected society. These too convenient justifications recall an earlier era when radio stations still believed that musicians should simply thank them for their songs being played at all.

The liability-based view of rights could aid in normalizing the deep data mining associated with "bioprospecting," whereby biomedical firms scour the globe for relatively rare genetic sequences—in plants, animals, and humans—to inform new drugs and treatments. While often condemned for exploiting and even desecrating native environments, the practice has also yielded some striking discoveries of general human benefit (Brown 2003, ch. 8). The proposal, then, is that bioprospectors would not seek prior permission to pursue specific scientific and commercial aims. Instead, they would make an upfront contribution to the people and areas from where they plan to extract materials, in anticipation of possible harm along the way. The law and economics school of jurisprudence calls this "ex ante compensation," which is presented as a fairer standard for contracts than "informed consent" where one wishes to maximize the freedom of the contracting parties under conditions of irreducible uncertainty, where consent can never be properly "informed" (Posner 1981, ch. 4). In practice, it would be imposed as a tax but function like an insurance policy that the bioprospectors would take out on behalf of the natives: The state would force the bioprospecting firms to insure the native populations against a variety of risks to which the natives might be exposed as a result of their activities. While the firms' profit margins would be cut, they would at least receive free roaming privileges in return. But perhaps most important of all, ex ante compensation converts the prima facie more powerful party, the bioprospectors, into de facto debtors with an incentive to act in a socially responsible manner from the outset of their activities.

This proactionary legal approach to exploring—indeed, exploiting—human resources equally provides an incentive for people to organize around shared information content, genetic or otherwise, in order to influence the agendas of bioprospectors and other intelligence-based firms (Fuller and Lipinska 2014, ch. 4). In many cases, these groups will consist of people who might not otherwise see themselves as having much in common. In the specific case of genetics, what Fuller and Lipinska (2014) call "hedgenetics" aims to ensure—in the spirit of royalties automatically paid for radio plays of copyrighted music—that those whose distinctive genetic makeup is relevant to biomedical innovation are compensated appropriately, even if the innovation is not successful. But also, and more importantly, hedgenetics would open up a legal vehicle for those with common genetic interests to advocate that more research be done on conditions related to their makeup.

A NEW PROBLEM OF JUSTICE FOR A NEW SENSE OF PERSONHOOD

The price of greater freedom is that others are free to access you, which means that you need to ensure that you benefit—or at least not harmed—by that newfound freedom that others have over you. The new problem of justice arises because the Internet allows everyone in principle to say what they wish, while allowing everyone else to hear what they say. Intuitive clarity about where "mine" ends and "thine" begins is disappearing, thereby challenging the liberal settlement. While in practice matters are not so symmetrical, there is no denying that the Internet enables people to develop means both to amplify what they can say and what they can hear. In this respect, surveillance is simply the complement of expression, if we take seriously the idea that people have a right to be heard (Kelly 2014). But how, then, do we ensure that someone does not say or hear "too much," such that they and/or others are prevented from freely expressing themselves in the future?

This question reminds us of the original value that started to be placed on privacy and publicity as clearly defined states of social being in the eighteenth century, namely, that some measure of hypocrisy is necessary for civil society to remain tolerant (Sennett 1977). Both Catholic inquisitions and Protestant witch hunts had presumed both a right and an ability on the part of religious authorities to penetrate the minds of others to determine whether what they say is really what they think. These practices reached their peak in the wars of religion that blighted the sixteenth- and seventeenth-century Judaeo-Christian world. Recall that the use of "atheist" as a term of abuse applied to those whose outward behavior conformed to Christian norms but who were nevertheless suspected as lacking a "true" connection with the deity, which in the world before Kant was understood as *both* moral and cognitive. (The trans/posthuman heir to this sensibility is arguably the concern that surrounds passing the Turing Test, especially in the manner presented in *Blade Runner*.) To be sure, "atheism" as a positive mode of social existence had been proposed by the ancient Greek skeptics on therapeutic grounds and was widely recommended of Jews as a survival strategy in the Christendom. But "tolerance" as a value upheld in modern law marks the first formal recognition of this previously demonized practice (Israel 2006, chs. 6–7). It was in this context that Spinoza appeared as a martyr for liberalism, as he was expelled by one (Jewish) religious community without seeking refuge in another one, let alone forming a new community (as most of the Protestants did).

However, the intuitions that historically supported the value of privacy are eroded in a world where one's identity is just as much associated with one's avatar or username as one's physical being. The liberal settlement, courtesy of John Locke, presumed that the mind is coextensive with the body in all its powers. This explained our unique personal identities as well as our difficulties in accessing the identities of others. However, Facebook is helping to dissolve the intuitions underwriting this ontology with its sliding scale of access settings, which implies that "privacy" is simply the absence of publicity. This in turn suggests that those who uphold privacy either, morally, have something bad to hide or, cognitively, are concealing their incompetence in the guise of rent—one increasingly common reading of "intellectual property" (Fuller 2002, chs. 1–2; Landes and Posner 2003). And so we seem to be back to Spinoza's world, where those who refuse to declare their faith are presumed to have no faith. Indeed, this world is back with a vengeance, since not even death can ensure that your digital remains will be left undisturbed by those wishing to exploit their full potential (Mayer-Schoenberger 2009).

The Bill of Rights—the first ten amendments to the U.S. Constitution—begins by asserting, first, the right to free expression and, second, the right to self-protection. The relationship between these two rights is clearest against the backdrop of the liberal settlement, whereby one mind = one body. To be sure, we can argue about whether these two rights are placed in the right order. Thus, perhaps the most influential account of rights in modern jurisprudence (Hohfeld 1919) regards self-protection as prior to self-expression, along lines that would please both Maslow and animal rights activists, whereas partisans of the law and economics movement see the two rights as orthogonal, given that many people appear happy to expose themselves to all sorts of harms in order to maximize self-expression (McKenzie and Tullock 2012, ch. 3). However, even the latter set of scholars, who generally endorse the U.S. Founding Fathers' ordering of the rights, presuppose a Lockean sense of individuality, in which the capacity of one's mind is ultimately limited by the capacities of one's body. However, it is just this exactness of fit that is now being challenged by the reconstruction of social life in the medium of information and communication technology. Even without much enhancing our bodies, our minds have become extended. As a result, privacy in its modern legal sense is effectively dead. We are in a transitional stage in human self-understanding, in which we come to learn how to be an *investor* in—as opposed to an *owner* of—who we are. This, in turn, will require a massive shift in our understanding of rights from a property to a liability basis.

BIBLIOGRAPHY

Barry, Brian. *Political Argument*. London: Routledge & Kegan Paul, 1965.

Brown, Michael F. *Who Owns Native Culture?* Cambridge, MA: Harvard University Press, 2003.

Calabresi, Guido and A. Douglas Melamed. "Property Rules, Liability Rules, and Inalienability: One View of the Cathedral." *Harvard Law Review* 85, no. 6 (1972): 1089–1128.

Cassirer, Ernst. *Substance and Function*. La Salle, IL: Open Court Press, [1910] 1923.

Clark, Andy. *Supersizing the Mind.* Cambridge, MA: MIT Press, 2008.
Elster, Jan. *Sour Grapes*. Cambridge, UK: Cambridge University Press, 1983.
Fuller, Steve. *Bounded Rationality in Law and Science*. PhD diss., University of Pittsburgh, 1985.
Fuller, Steve. *Knowledge Management Foundations*. Woburn, MA: Butterworth-Heinemann, 2002.
Fuller, Steve. "Precautionary and Proactionary as the New Right and the New Left of the Twenty-First Century Ideological Spectrum." *International Journal of Politics, Culture and Society* 25, no. 4 (2012): 157–74.
Fuller, Steve and Veronika Lipinska. *The Proactionary Imperative: A Foundation for Transhumanism*. London: Palgrave, 2014.
Galbraith, John Kenneth. *The Affluent Society*. Boston: Houghton Mifflin, 1958.
Hare, Rom M. "Rawls' Theory of Justice." *Philosophical Quarterly* 23 (1973): 144–55.
Hohfeld, Wesley N. *Fundamental Legal Conceptions*. New Haven: Yale University Press, 1919.
Israel, Jonathan. *Enlightenment Contested: Philosophy, Modernity and the Emancipation of Man, 1670–1752*. Oxford: Oxford University Press, 2006.
Kelly, Kevin. "Why You Should Embrace Surveillance, Not Fight It." *Wired*, March 10, 2014.
Kurzman, Charles. *Democracy Denied, 1905–1915: Intellectuals and the Fate of Democracy*. Cambridge, MA: Harvard University Press, 2008.
Landes, William and Richard Posner. *The Economic Structure of Intellectual Property Law*. Cambridge, MA: Harvard University Press, 2003.
Lessig, Lawrence. *The Future of Ideas*. New York: Random House, 2001.
Maslow, Abraham. *The Farther Reaches of Human Nature*. New York: Viking, 1971.
Mason, Paul. *Why It's Kicking Off Everywhere: The New Global Revolutions*. London: Verso, 2012.
Mayer-Schoenberger, Viktor. *Delete: The Virtue of Forgetting in the Digital Age*. Princeton: Princeton University Press, 2009.
McKenzie, Richard and Gordon Tullock. *The New World of Economics*, 6th ed. Berlin: Springer, [1975] 2012.
Morozov, Evgeny. *The Net Delusion*. New York: Public Affairs, 2011.
Morozov, Evgeny. *To Save Everything, Click Here*. New York: Public Affairs, 2013.
Pentland, Alex. *Honest Signals*. Cambridge, MA: MIT Press, 2008.
Polanyi, Karl. *The Great Transformation*. Boston: Beacon Press, 1944.
Posner, Richard. *The Economics of Justice*. Cambridge, MA: Harvard University Press, 1981.
Quine, W. V. O. *From a Logical Point of View*. New York: Harper & Row, 1953.
Rawls, John. *A Theory of Justice*. Cambridge, MA: Harvard University Press, 1971.
Sennett, Richard. *The Fall of Public Man*. New York: Alfred Knopf, 1977.
Stinchcombe, Arthur. *Information and Organizations*. Berkeley: University of California Press, 1990.

Epilogue

The Future of Social Epistemology: A Collective Vision comes out of a sustained experiment in effecting social epistemology as a collective practice. We ask for your help in continuing this experiment. If you take the social formation of knowledge and belief as an object of study, or wish to reply to the arguments and ideas in this book, or to any of the books published in the *Collective Studies in Knowledge and Society* series, please accept our invitation to engage with the Social Epistemology Review and Reply Collective. To learn how to participate, go to: social-epistemology.com/about/.

We realize the future together.

Index

About the Contributors

Guy Axtell is an associate professor of philosophy in the Department of Philosophy and Religious Studies at Radford University. He is most recently author of *Objectivity* in the Polity Press Key Concepts in Philosophy series (2015). His research elaborates intersections between social, feminist, and virtue epistemologies, while suggesting new overlapping research interests between science studies, sociology, and his own field of philosophy.

Laura Cabrera is a postdoctoral research fellow at the National Core for Neuroethics at the University of British Columbia. She has been conducting research exploring the attitudes of the general public toward enhancing interventions, and the normative implications of using neurotechnologies for nonmedical purposes. Her career goal is to pursue interdisciplinary neuroethics scholarship, provide active leadership, and train and mentor future leaders in the field.

James H. Collier is an associate professor in the Department of science and technology in Society at Virginia Tech. He is the executive editor of *Social Epistemology*, the founder and acting editor of the *Social Epistemology Review and Reply Collective*, and the series editor of "Collective Studies in Knowledge and Society." Most recently, he edited *On Twenty-Five Years of Social Epistemology: A Way Forward* (Routledge, 2013).

Emma Craddock was awarded an Economic and Social Research Council 1+3 studentship at the University of Nottingham in 2011. She completed the masters in social research methods in 2012 with distinction and began her PhD in September 2012, researching individuals' meanings and experiences of political activism against austerity in Nottingham, UK.

Fred D'Agostino is professor of humanities and president of the Academic Board at the University of Queensland. He coedited the *Routledge Companion to Social and Political Philosophy.* His most recent book is *Naturalizing Epistemology.* He is working on the disciplines. He is a fellow of the Australian Academy of Humanities.

William Davis is a doctoral candidate in science and technology studies at Virginia Tech. His recent work argues that the ethical, legal, political, and social implications of emerging technologies, and their concomitant values, must become part of general public discussions if we wish to exercise influence on the development and deployment of technologies that will impact our future lives and societies.

Susan Dieleman is currently assistant professor of philosophy (with term) at the University of Saskatchewan in Saskatoon, Canada. Her research projects involve bringing pragmatist feminist tools to bear on issues of (epistemic) justice and (deliberative) democracy. Her work has been published in *Hypatia, Pragmatism Today, Social Epistemology, Social Philosophy Today, Administration and Society*, and *The Pluralist.*

Martin Evenden is an assistant professor in the Department of English at National Taichung University of Education.

Steve Fuller is Auguste Comte Professor of Social Epistemology in the Department of Sociology at the University of Warwick, UK. Originally trained in history and philosophy of science, Fuller is best known for his foundational work in the field of "social epistemology," which is the name of a quarterly journal that he founded in 1987 as well as the first of his more than twenty books. His most recent work has been concerned with future of "humanity" as both a being and a concept.

Inanna Hamati-Ataya is reader in international politics at Aberystwyth University, UK. Her research lies at the intersection of the fields of international politics, social-political theory, and the sociology of knowledge and science. She is currently coeditor of the *Worlding beyond the West* book series at Routledge, and coeditor of the IR Theory Section of the *International Studies Encyclopedia* published by Wiley-Blackwell.

Eric Kerr is joint postdoctoral fellow at the Asia Research Institute (Science, Technology, and Society Cluster) and a fellow at Tembusu College, National University of Singapore. His research interests are primarily in the philosophy of technology and social epistemology, with a special focus on petroleum engineering. Eric completed his PhD in philosophy at the University of Edinburgh in 2013.

Joan Leach, University of Queensland, Australia, is currently the president of Australian Science Communicators and chair of the National Committee for HPS at the Australian Academy of Science. Her recent research explores the globalization of popular science since the 1960s. From January 1, 2016, she will be the director of the Centre for Public Awareness of Science at the ANU.

Veronika Lipinska is a lawyer and a lecturer at the European School of Law and Administration in London. She is a coauthor (with Steve Fuller) of *The Proactionary Imperative* and a contributor to many transhumanist publications, including articles in the Wiley-Blackwell *Encyclopaedia of Social Theory* and the *Encyclopedia of Ethics, Science, Technology and Engineering*-2. Veronika is also a member of the Lifeboat Foundation.

Jonathan Matheson is an associate professor in the Department of Philosophy and Religious Studies at the University of North Florida. He has written numerous articles in epistemology, is a coeditor of *The Ethics of Belief: Individual and Social* (Oxford University Press), and is the author of *The Epistemic Significance of Disagreement* (Palgrave).

Fabien Medvecky is a lecturer in science communication at the University of Otago's Centre for Science Communication. With degrees in philosophy of science and environmental economics, he previously lectured at the University of Queensland's program in science communication. Fabien has published on ethics, economics, and science communication, and is currently working on the sociality of economics as a discipline.

María G. Navarro is a postdoctoral researcher at the Spanish National Research Council and an affiliated researcher of the European Centre for Soft Computing. Her research interests are in social epistemology; the intersection of argumentation theories with theory of controversies and hermeneutics; and philosophy of history.

Stephen Norrie, since finishing his PhD on the foundations of Marxism, has been working outside academia, on the idea of a critique of philosophy. Based in central England, he has previously written on the relation of the idea of critical theory to the social theory of high culture, with reference to the work of György Márkus.

Philip R. Olson is assistant professor in the Department of Science and Technology in Society at Virginia Tech. Drawing upon his philosophical background in ethics and epistemology, Phil's science and technology studies (STS) research takes place at the intersections of bioethics, technology ethics, conceptions of the body (living and dead), social epistemology, and women's and gender.

Melissa Orozco is a third-year PhD student in the program of philosophy of science at the National Autonomous University of Mexico. Her research focuses on the conception of normativity in social epistemology and the gaps this new field still represents for discussions related to the study of science and technology.

Victoria Peake studied BA (Hons), sociology, at the University of Warwick during which her research focused on human enhancement technologies and the changing notion of what it means to be "human." She is particularly interested in the ways art and bioart can shape and reflect technological and scientific advances. Victoria is currently working as a safeguarding officer for Swale Borough Council in Kent.

David Budtz Pedersen is an associate professor and co-director of the Humanomics Research Centre, University of Copenhagen. His research focuses on social epistemology and science and innovation policy. He holds PhD, MA, and BA degrees in philosophy and science policy studies. David has about seventy-five entries on his list of publications ranging from research papers to monographs, edited volumes, reports, columns, and newspaper articles.

Patrick J. Reider is an assistant professor at the University of Pittsburgh at Greensburg. Currently, he is an editor and contributing author for two books: *Wilfrid Sellars, Idealism and Realism: Understanding Psychological Nominalism* (Bloomsbury Publishing) and *Social Epistemology and Epistemic Agency: Decentralizing the Epistemic Agent* (Rowan and Littlefield International). He is also the special issue editor for the *Social Epistemology Review and Reply Collective.*

Francis Remedios is an independent scholar with a 2003 book *Legitimizing Scientific Knowledge: An Introduction to Steve Fuller's Social Epistemology.* He published several papers on social epistemology and serves on the editorial board of the journal *Social Epistemology.* His current research is on social epistemology and transhumanism.

Adam Riggio has left the search for full-time work as university faculty to pursue a career in communications, activism, and writing. He believes that he can achieve more for human progress outside the university system. He completed his PhD in philosophy at McMaster University in 2012; and Palgrave Macmillan published the final product of his dissertation research in 2015 as *Ecology, Ethics, and the Future of Humanity.*

Diana Rishani is a student currently pursuing her master's in anthropology at the University of Ottawa, Canada. She holds a double degree in philosophy and anthropology from the American University of Beirut, Lebanon. She aims later on to pursue a PhD in notions relating to the manifestation of power relations within society and further contribute to the field.

Gregory Sandstrom is a lecturer at the European Humanities University in Vilnius, Lithuania. He is author of the books *Human Extension: An Alternative to Evolutionism, Creationism and Intelligent Design* (Palgrave Macmillan, 2014) and *Evolution's Puzzle in Social Sciences and Humanities* (EHU Press, 2015). His chapter was written during a postdoctoral fellowship with the Lithuanian Research Council.

Elisabeth Simbürger (PhD, MA sociology, University of Warwick) works as a senior lecturer at the Department of Sociology at the University of Valparaíso, Chile. Her work focuses on the politics of knowledge in higher education with an emphasis on the epistemological transformation of sociology as a discipline as a result of historical, political, and economic change.

Miika Vähämaa graduated in 2015 from the University of Helsinki. His PhD dissertation in social psychology focused on the epistemic dimension of social groups and initiated a synthesis of social forces that also act as epistemic standards. Dr. Vähämaa has been a Fulbright scholar at the University of North Carolina at Chapel Hill and he now teaches at the Open University of the University of Helsinki.

Mark Douglas West is a professor of mass communication at the University of North Carolina at Asheville. After serving as a reporter, he worked in opinion research before completing his PhD; his dissertation, upon the relationship between public opinion, policy, and press coverage of the Vietnam War, won two international awards. Dr. West has published five books on varied subjects.

Emilie Whitaker is a lecturer in social science at Cardiff University. She is an ethnographer operating at the intersection between sociology and social policy. Her work centers on the sociology of the life course with a specific interest in policies, cultures, and practices of care and welfare. She is particularly interested in the interrelational aspects of transhumanist futures.

Pedro Saez Williams is a final year doctoral researcher in the department of sociology at the University of Warwick. His research is concerned with the dimensionality of normativity and social authority (both cognitive and political), specifically focused on developing an account of normativity compatible with naturalism. He holds an MPhil in sociology from the University of Cambridge.

www.ingramcontent.com/pod-product-compliance
Ingram Content Group UK Ltd.
Pitfield, Milton Keynes, MK11 3LW, UK
UKHW041126071026
15587UKWH00009B/254

9 781783 482658